Old Protein, New Medicine - Brain-Derived Neurotrophic Factor

Edited by Oytun Erbaş and İlknur Altuntaş

Published in London, United Kingdom

Old Protein, New Medicine - Brain-Derived Neurotrophic Factor
http://dx.doi.org/10.5772/intechopen.111201
Edited by Oytun Erbaş and İlknur Altuntaş

Contributors
Ejder Saylav Bora, Berzah Güneş, Lora Koenhemsi, Oytun Erbaş, Rastegar Hoseini, Zahra Hoseini, Elahe Bahmani, Mehmet Burak Yalçın, Mumin Alper Erdogan, Sudhiranjan Gupta, Rakeshwar S. Guleria

First published in London, United Kingdom, 2023 by IntechOpen
IntechOpen is the global imprint of INTECHOPEN LIMITED, registered in England and Wales, registration number: 11086078, 5 Princes Gate Court, London, SW7 2QJ, United Kingdom
Printed in Croatia

British Library Cataloguing-in-Publication Data
A catalogue record for this book is available from the British Library

Additional hard and PDF copies can be obtained from orders@intechopen.com

Old Protein, New Medicine - Brain-Derived Neurotrophic Factor
Edited by Oytun Erbaş and İlknur Altuntaş
p. cm.

This title is part of the Biochemistry Book Series, Volume 47
Topic: Proteomics
Series Editor: Miroslav Blumenberg
Topic Editors: Paolo Iadarola and Simona Viglio

Print ISBN 978-1-83769-312-2
Online ISBN 978-1-83769-313-9
eBook (PDF) ISBN 978-1-83769-314-6
ISSN 2632-0983

For EU product safety concerns:
IN TECH d.o.o., Prolaz Marije Krucifikse Kozulić 3, 51000 Rijeka, Croatia,
info@intechopen.com or visit our website at intechopen.com.

IntechOpen Book Series
Biochemistry
Volume 47

Aims and Scope of the Series

Biochemistry, the study of chemical transformations occurring within living organisms, impacts all of the life sciences, from molecular crystallography and genetics, to ecology, medicine and population biology. Biochemistry studies macromolecules - proteins, nucleic acids, carbohydrates and lipids –their building blocks, structures, functions and interactions. Much of biochemistry is devoted to enzymes, proteins that catalyze chemical reactions, enzyme structures, mechanisms of action and their roles within cells. Biochemistry also studies small signaling molecules, coenzymes, inhibitors, vitamins and hormones, which play roles in the life process. Biochemical experimentation, besides coopting the methods of classical chemistry, e.g., chromatography, adopted new techniques, e.g., X-ray diffraction, electron microscopy, NMR, radioisotopes, and developed sophisticated microbial genetic tools, e.g., auxotroph mutants and their revertants, fermentation, etc. More recently, biochemistry embraced the 'big data' omics systems. Initial biochemical studies have been exclusively analytic: dissecting, purifying and examining individual components of a biological system; in exemplary words of Efraim Racker, (1913 –1991) "Don't waste clean thinking on dirty enzymes." Today, however, biochemistry is becoming more agglomerative and comprehensive, setting out to integrate and describe fully a particular biological system. The 'big data' metabolomics can define the complement of small molecules, e.g., in a soil or biofilm sample; proteomics can distinguish all the proteins comprising e.g., serum; metagenomics can identify all the genes in a complex environment e.g., the bovine rumen.
This Biochemistry Series will address both the current research on biomolecules, and the emerging trends with great promise.

Meet the Series Editor

Miroslav Blumenberg, Ph.D., was born in Subotica and received his BSc in Belgrade, Yugoslavia. He completed his Ph.D. at MIT in Organic Chemistry; he followed up his Ph.D. with two postdoctoral study periods at Stanford University. Since 1983, he has been a faculty member of the RO Perelman Department of Dermatology, NYU School of Medicine, where he is codirector of a training grant in cutaneous biology. Dr. Blumenberg's research is focused on the epidermis, expression of keratin genes, transcription profiling, keratinocyte differentiation, inflammatory diseases and cancers, and most recently the effects of the microbiome on the skin. He has published more than 100 peer-reviewed research articles and graduated numerous Ph.D. and postdoctoral students.

Meet the Volume Editors

Prof. Oytun Erbaş is an expert in physiology and experimental medicine. His main field of experience is pharmaceutical R&D studies, neuroscience, biological psychology, and pathophysiology. He currently works in the Department of Pathophysiology at various universities. He is the founder of Experimental Medicine R&D labs in Illinois, USA, and Istanbul-Türkiye. For his experimental work, he received the 2013 Politzer Prize from the International Society of Otologic Surgery in the field of otolaryngology and the 2013 Jacques Duparc Award from the European Society of Orthopedics and Traumatology. His research interests include inflammation and psychiatric relationships, oxytocin effects, mechanisms and treatment of epilepsy, neurodevelopmental disorders (autism) and treatment, and neuroprotective drug trials in neuropathy.

İlknur Altuntaş, Ph.D., is a molecular biologist. Her main field of experience is oxidative stress, oxidative DNA damage, and neuroscience. Her scientific interests include inflammation and psychiatric relationships, neurodevelopmental disorders, and treatment.

Contents

Preface

As a member of the neurotrophin family of proteins, brain-derived neurotrophic factor (BDNF) plays a vital role in maintaining optimal brain function, encompassing functions such as synaptic plasticity, cellular differentiation, learning processes, and the preservation of nerve cells. Neural plasticity, the nervous system's capability to adapt to varying environmental conditions, involves a diverse array of structural and functional mechanisms. Due to its robust neuroprotective properties and recently uncovered anti-inflammatory and anti-apoptotic attributes observed both in laboratory settings and in living organisms, BDNF has long been suggested as a potential preventative measure against neurodegeneration.

BDNF has emerged as a significant player in the pathophysiology of numerous psychiatric disorders, including depression, anxiety, schizophrenia, and bipolar disorder. Empirical evidence has consistently indicated that individuals afflicted with these conditions frequently exhibit lower BDNF levels in both their blood and brains.

The precise role of BDNF in psychiatric disorders remains an active area of research. However, scientists hypothesize that BDNF may contribute to several fundamental symptoms observed in these disorders, such as mood fluctuations, cognitive impairments, and social withdrawal. In addition to these considerations, BDNF has some potential additional benefits in addressing psychiatric disorders. *Enhancing mood*: BDNF can stimulate the release of mood-regulating neurotransmitters, such as serotonin and dopamine. *Reducing anxiety and stress*: BDNF has the capacity to modulate the stress response, contributing to a reduction in anxiety symptoms. *Enhancing cognitive function*: BDNF may enhance memory, attention, and executive function, functions that can be compromised in individuals with psychiatric disorders. *Facilitating social interaction*: BDNF can bolster the development and operation of brain circuits associated with social interaction.

BDNF has emerged as a significant factor in the understanding of autism spectrum disorder (ASD) pathophysiology. Studies have consistently indicated that individuals with ASD often exhibit lower levels of BDNF in both their blood and brains. Moreover, there is evidence of BDNF gene variants being linked to an increased risk of ASD.

The precise role of BDNF in ASD remains a subject of ongoing research. Nonetheless, scientists posit that BDNF may contribute to some of the central symptoms associated with ASD, including challenges in social communication and the presence of narrow interests.

BDNF stands as a promising target for therapeutic interventions in ASD. Current efforts by researchers are focused on the development of novel drugs and therapies designed to elevate BDNF levels or enhance BDNF signaling. It is crucial to pinpoint the specific BDNF pathways that play a role in ASD and create biomarkers capable of predicting treatment responses.

Beyond these considerations, BDNF has additional benefits for individuals with ASD. *Enhancing social communication skills*: BDNF can support the development and functioning of brain circuits crucial for social communication. *Mitigating restricted interests and repetitive behaviors*: BDNF may assist in diversifying interests and behaviors. *Improving cognitive function*: BDNF has the potential to enhance memory, attention, and executive function. *Alleviating anxiety and depression*: BDNF exhibits mood-boosting effects and may help reduce symptoms of anxiety and depression. BDNF represents a promising focal point for therapeutic endeavors in the context of ASD.

Ketamine is a dissociative anesthetic renowned for its rapid and enduring antidepressant effects. Researchers are also exploring its potential as a treatment for a range of psychiatric and neurological conditions, including anxiety, bipolar disorder, post-traumatic stress disorder (PTSD), and chronic pain. The therapeutic impact of ketamine is believed to be orchestrated through various mechanisms, including its influence on the N-methyl-D-aspartate receptor, modulation of glutamate signaling, and the promotion of BDNF production.

BDNF is believed to be instrumental in the formation of new synapses, vital for learning and memory. This might elucidate why ketamine displays efficacy in treating depression, a condition often accompanied by a reduction in synaptic connections. Furthermore, BDNF is thought to participate in neurogenesis, the process of generating new neurons. Neurogenesis is compromised in several psychiatric disorders, including depression, anxiety, and PTSD. Ketamine has demonstrated the potential to stimulate neurogenesis in animal models of these disorders, hinting at its potential utility in human treatments.

Collectively, evidence points to BDNF as a crucial mediator of ketamine's therapeutic effects. Ketamine-induced elevations in BDNF levels may underlie many of its therapeutic benefits, encompassing its antidepressant, anxiolytic, and analgesic properties.

The significance of BDNF extends to the realm of spinal cord injury (SCI), where it is intricately involved in the pathophysiological processes. BDNF exhibits a diverse range of neuromodulatory effects within the spinal cord. *Enhancing synaptic plasticity*: BDNF actively fosters the formation and maintenance of synapses, crucial for the preservation of learning and memory functions. *Promoting axonal regeneration*: BDNF acts as a stimulant for axonal growth, a vital process in the repair of nervous system damage. *Safeguarding neurons from demise*: BDNF provides vital support for neuronal survival and shields them from potential injury.

In preclinical models of SCI, the administration of BDNF has consistently demonstrated improvements in locomotor function, sensory perception, as well as bladder and bowel function. Moreover, BDNF has showcased its capacity to encourage axonal regeneration and protect neurons from degeneration.

Sepsis is a life-threatening condition that arises when the body's response to an infection inadvertently damages its own tissues and organs, often resulting in organ dysfunction and mortality. It stands as a leading cause of death among critically ill patients.

BDNF is a pivotal protein involved in the development and maintenance of the nervous system, with additional roles in immune system regulation. In sepsis, BDNF levels in both the blood and brain frequently experience a decline. This reduction is associated with several adverse consequences, including organ dysfunction, cognitive impairment, and fatal outcomes.

BDNF holds substantial promise as a therapeutic target for sepsis. Ongoing research endeavors are focused on the development of novel drugs and therapies designed to elevate BDNF levels or enhance BDNF signaling.

In addition to these considerations, BDNF confers additional benefits in addressing sepsis. *Mitigating inflammation*: BDNF's anti-inflammatory properties could potentially help alleviate the organ damage characteristic of sepsis. *Shielding neurons from demise*: BDNF has demonstrated the ability to safeguard neurons from the harm associated with sepsis-induced encephalopathy. *Enhancing cognitive function*: BDNF may aid in the improvement of cognitive function, a facet often impaired in survivors of sepsis. *Facilitating immune recovery*: BDNF may contribute to the restoration of immune system functionality, which is frequently dysregulated in sepsis.

Both exercise and vitamin D have demonstrated the ability to elevate BDNF levels within the brain. Exercise is believed to achieve this by boosting the production of insulin-like growth factor 1, a hormone known to stimulate BDNF production. On the other hand, vitamin D is thought to enhance BDNF levels by binding to the vitamin D receptor in the brain and activating genes associated with BDNF production.

Emerging evidence suggests that the combination of exercise and vitamin D may exert a synergistic influence on BDNF levels.

The combined impact of exercise and vitamin D on BDNF levels holds the potential for numerous advantages in maintaining brain health. Additionally, BDNF plays a vital role in shielding neurons from damage, potentially reducing the risk of developing neurodegenerative conditions such as Alzheimer's disease and Parkinson's disease. The current evidence points to the combination of exercise and vitamin D as a promising strategy for heightening BDNF levels and enhancing brain health.

This book is a comprehensive source of knowledge on BDNF and its neuroprotective functions. The chapters within offer insights into recent advancements, molecular principles, and innovative therapeutic strategies targeting neurodegenerative disorders and brain health.

Oytun Erbaş
ERBAS Institute of Experimental Medicine,
Gebze-Kocaeli, Türkiye

ERBAS Institute of Experimental Medicine,
Illinois, USA

İlknur Altuntaş
ERBAS Institute of Experimental Medicine,
Gebze-Kocaeli, Türkiye

Section 1

Brain-Derived Neurotrophic Factor and Psychiatric Disorders

Chapter 1

The Role of Brain-Derived Neurotrophic Factor in Psychiatric Disorders

Sudhiranjan Gupta and Rakeshwar S. Guleria

Abstract

Brain derived neurotrophic factor (BDNF) is one of the most extensively studied and widespread growth factors in the brain. BDNF and its receptors are the critical factors having multipotent impact on the central nervous system (CNS). The biological function of BDNF primarily mediated by two receptors, tropomyosin receptor kinase B (TrkB) receptor and p75 neurotrophin receptor. BDNF contributes a pivotal role in neuronal and glial development, modulation and maintaining overall synaptic plasticity of the brain; therefore, widely involved in psychiatric diseases. Current hypotheses indicates that abnormal BDNF level, a vital condition for psychiatric and neurodegeneration diseases are mainly due to the disruption of the BDNF-associated signaling cascades. It is, therefore, crucial to understand how BDNF coordinate the psychiatric diseases in the brain. This review begins with the history of BDNF and its biology in brain homeostasis and focuses on several aspects of BDNF signaling. In addition, the review addresses the impact of BDNF level in diverse neuropsychiatric disorders including major depressive disorder, schizophrenia, bipolar disorder, post-traumatic stress disorder and, possible biological mechanisms of BDNF that may shed new insight for future therapeutic use and drug development.

Keywords: BDNF, inflammation, brain homeostasis, brain plasticity, psychiatric disorders

1. Introduction

Brain-derived neurotrophic factor (BDNF) is a neurotrophin classified as dimeric polypeptide regulating a wide array of neuronal activities including but not limited to neurogenesis, neuronal growth, differentiation, excitability, and plasticity. BDNF was originally identified by Barde et al. [1] as a factor from cultured embryonic chick which showed survival of sensory neurons. Soon after its discovery, BDNF was recognized and laid a foundation for neuronal plasticity in the adult brain and further observed its' pivotal role in neuronal activity [2–4]. Subsequently, BDNF was considered for antidepressant treatments therapy as it was shown that neurotrophins promoted the growth and helped in maturation of neurons [5–7]. Interestingly, injection of BDNF in the hippocampus elicited antidepressant-like effects in rodents led to advocate a critical role for BDNF in the setting formulating antidepressant drugs [8–10]. The line of research identified BDNF and its cognate receptor tropomyosin receptor kinase (TrkB, neurotrophic tyrosine kinase receptor, NTRK2) in the hippocampus and cortex suggested antidepressant drug action into neuronal plasticity [11].

BDNF contributed a key role in the development of the nervous system by regulating neuronal development, growth, differentiation, neurogenesis, synaptogenesis, and synaptic plasticity [12–14]. Moreover, neurodegenerative, and neuropsychiatric diseases appear to be linked with insufficient BDNF level leading to the defects in synaptic plasticity [15, 16]. As a result, strategies to increase the BDNF level in circulation was advocated for therapy in neurological diseases.

This article reviews the current understanding and future directions in BDNF-related research in the central nervous system, with an emphasis on the possible therapeutic application of BDNF in modifying fundamental processes underlying neural disease.

2. BDNF, a neurotrophin family member: synthesis, secretion and function

Nearly three decades earlier discovery of nerve growth factor (NGF) by Rita Levy-Montalcini [17], prompted Yves-Alain Barde searched for a growth factor with similar properties and function like NGF in neurons. The study culminated into a purified protein from pig brain named BDNF [1]. Later, amino acid sequence revealed that BDNF shared a significant homology with NGF along with other members like neurotrophin 3 and neurotrophin 4, together constitute a conserved neurotrophin family [18].

Synthesis and maturation of BDNF is a multistage process, involving formation of several precursor isoforms. BDNF is initially synthesized in the Golgi after cleaving the signal sequence from pre region as a precursor form (pro-BDNF) containing 129 amino acids N-terminal prodomain and a 118 amino acids C-terminal mature domain [19]. The mature domain forms a cysteine knot structure, leading to non-covalent dimerization of the mature domains [20]. When the prodomain is cleaved from intact pro-BDNF, through the actions of proconvertase at a conserved RVRR sequence, the dimeric mature domains are released, and are called mature BDNF, or simply BDNF [21]. Secretion of m-BDNF and pro-BDNF into the extracellular space enables their physiological action (see the diagram, **Figure 1**).

In neuronal cells, both pro-BDNF and m-BDNF are released following cellular membrane depolarization and maintained a dynamic balance [22–24]. Both isoforms are important in neuronal function in the brain, but mature-BDNF (m-BDNF) appeared to offer neurogenesis, neuroprotection, synaptic plasticity, and synaptic function in neurons [25, 26]. The m-BDNF is axonally delivered into axon vesical terminals followed by the secretion into axonal cleft [22]. Mechanistically, BDNF requires to bind its' partner/receptor, Tr, located both pre- and post-synaptic membrane, to complete its function. BDNF is highly conservative and is expressed as a single gene, *Bdnf* transcript and is dynamically regulated and showed cell-specific neural activity. The human *Bdnf* gene, a ~ 70 kb, is in the chromosome 11 consisting of 11 exons (I-IX along with Vh and VIIIh) in the 5′ end and 9 promoters in tissues and brain regions [27, 28]. Apart from the above-mentioned BDNF isoforms, the function of BDNF is potentially affected by single nucleotide polymorphism of methionine (Met) to valine (Val) substitution at 66th position of *Bdnf* gene.

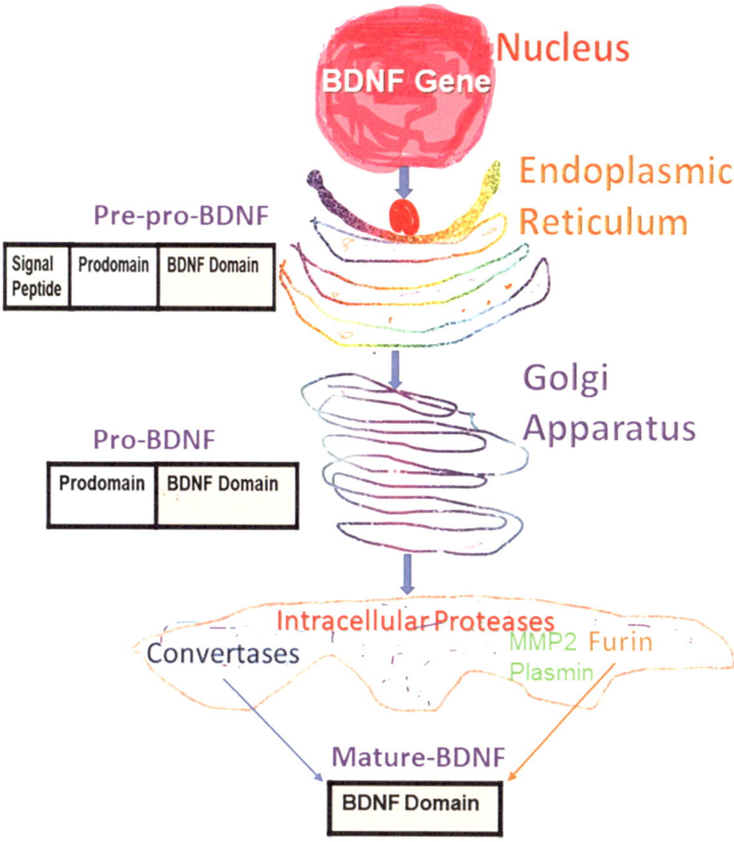

Figure 1.
Schematic presentation of synthesis and maturation of BDNF. In the intracellular pathway, the pre-pro-BDNF precursor molecule is produced in the endoplasmic reticulum and transported to the Golgi apparatus. During intracellular cleavage, the pre-region is removed, resulting in formation of immature isoform of BDNF called pro-BDNF. Finally, the pro-domain is removed and the mature isoform of BDNF, m-BDNF is produced. The cleavage process is mediated by intracellular proteases, convertases, and furin resulting the release of both pro-BDNF and m-BDNF isoforms into the extracellular space. Here, it is further processed by metalloproteinases 2 and 9 (MMP2 and MMP9), and plasmin.

Considering BDNF neuronal function, it is more appreciated as differentiation factor than survival neurotrophin [29, 30]. In addition to synaptic transmission, BDNF elicits long-term potentiation in hippocampus and modulate neuronal circuit function [31]. Moreover, changes in BDNF level in rodent models demonstrated aberrant function in hippocampal regions, including impaired memory, aggression, and hyperphagia [32].

3. BDNF receptors and intracellular signaling

BDNF signals are mediated by TrkB receptor and p75 neurotrophin receptor. BDNF binds with high affinity with TrkB, a tyrosine kinase receptor family, and the p75 neurotrophin receptor (p75 NTR), a member of the tumor necrosis factor (TNF) receptor family and low with p75 receptor. The TrkB is widely expressed in brain including cortex, hippocampus and in spinal cord nuclei [33]. It is noted that the mature BDNF binds to TrkB whereas pro-BDNF binds to p75NTR. The pro-BDNF/p75NTR signaling primarily promoting synaptic elimination by activating c-Jun N-terminal Kinase (JNK) pathway and triggers apoptosis. Other family members of Trk are TrkA which is specific to NGF [34] and TrkC which binds other neurotrophins [35]. This review will focus TrkB and its' signaling.

Activation of BDNF begins by binding to TrkB, and dimerizing and activating intrinsic kinase cascade before going to autophosphorylation. The BDNF/TrkB complex gets internalized into the neuron and serves as a docking site for diverse signaling platforms, protein phosphorylation and secondary signaling events [36, 37]. Next, the binding of BDNF to TrkB receptor, BDNF/TrkB in complex, leads to phosphorylation and translocation of TrkB into cellular membrane lipid rafts, and activating diverse important intracellular signaling cascades for performing cellular functions that include mitogen-activated protein kinase/extracellular signal-related kinase (MAPK/ERK), guanosine triphosphate hydrolases (GTP-ases) of the Ras homolog (Rho) gene and phospholipase C-γ (PLC-γ), phosphatidylinositol 3-kinase/protein kinase B (PI3K/AKT) pathways [38–41]. It is evidenced that PI3K/AKT pathway contributed to synaptic plasticity and cell survival or antiapoptotic activity response by modulating N-methyl-D-aspartate receptor (NMDAR) [40, 42]. Furthermore, BDNF-dependent neuroprotection is mediated via NMDAR/Ca^{2+} synaptic signaling resulting eliminating glutamatergic toxicity and preventing mitochondrial dysfunction and cellular apoptosis [43, 44]. The PLC g-dependent signaling triggers Ca^{2+}-calmodulin-dependent protein kinase (CAMK) and protein kinase C (PKC) to stimulate actin/microtubule synthesis and enhance synaptic plasticity and neuronal fiber growth [40, 45, 46]. The MAPK/Ras signaling regulates neural differentiation [45]. The ERK ½ and cAMP response element-binding protein (CREB) activation are necessary for cytoskeleton protein synthesis for dendritic growth and branching [40, 47]. In summary, the participation of BDNF in several physiological roles in the brain involves different signaling and is pivotal in maintaining a dynamic balance between the stimulus and its' function. A diagrammatic presentation of BDNF receptor and signaling is shown in **Figure 2**.

4. BDNF and brain homeostasis

Homeostasis is a fundamental process and equates to a dynamic balance between interdependent element and the physiological function in the organ of a living system. BDNF plays a significant role in neuronal plasticity in the central and peripheral nervous system [48]. BDNF is expressed throughout the development and adulthood in neurons of the brain and contributing a critical role in many physiological functions. One of the functions is energy homeostasis in the hypothalamus. Energy homeostasis is a complex gets interaction between the brain and peripheral tissues. Neuronal circuitry in the hypothalamus and hindbrain contributes a critical role in orchestrating the peripheral signals associated with energy storage by regulating nutrient intake and energy expenditure. BDNF is synthesized in several regions of hypothalamus including ventromedial hypothalamic nucleus (VMH), the dorsomedial hypothalamic nucleus (DMH), the paraventricular nucleus (PVH) and the lateral hypothalamic area (LH) [49, 50]. In particular, the energy balance is reported to be in the PVH region as evidenced by loss of body weight by injecting BDNF in this region [51]. The report showed that decrease in food intake resulted in increased resting metabolic rate, partly due to upregulation of uncoupling protein 1 (UPC1)

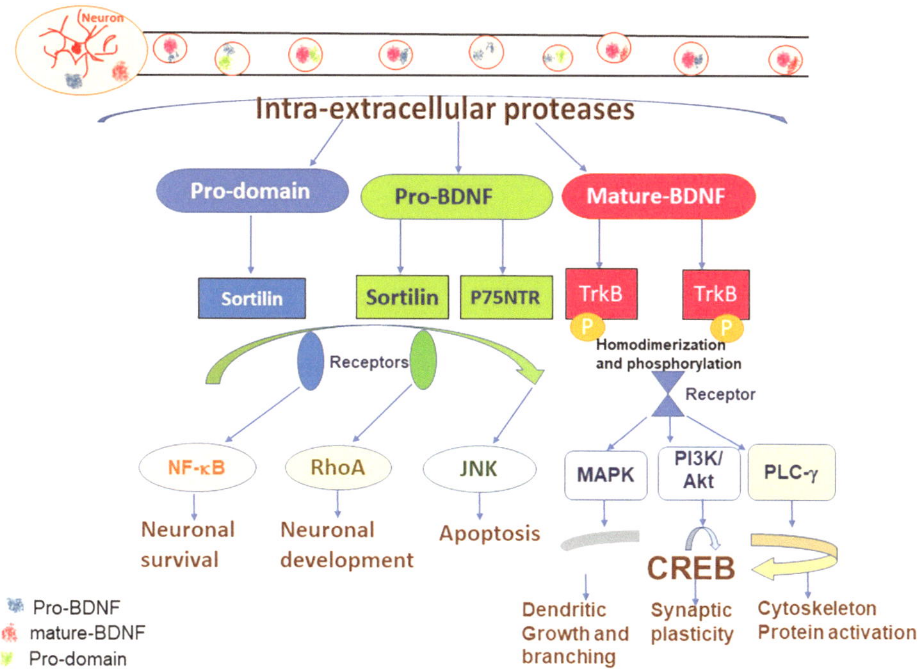

Figure 2.
BDNF signaling cascade. The BDNF is primarily transcribed as a precursor (pro-BDNF) which is later cleaved intra or extracellularly into mBDNF. The pro-BDNF exhibits affinity to sortilin and p75NTR receptors leading to the activation of nuclear factor κB (NF-κB), RhoA and JNK signaling pathways. The functional outcome of theses pathways includes neuronal survival, development, and apoptosis. The mBDNF showed highest affinity towards TrkB receptors. The mBDNF/TrkB complex triggers signaling pathways linked to phosphatidylinositol 3-kinase (PI3K), phospholipase C-γ (PLC-γ) and mitogen activated protein kinase (MAPK) via CREB. The pathways are involved in dendritic growth and branching, synaptic plasticity, and cytoskeleton protein activation.

in the brown adipose tissue [51]. Hypothalamic injection of BDNF promotes switching white adipose tissue to brown adipose tissue via sympathetic neuron activation and accelerates UCP-1 expression [52, 53]. This is an example of the role of BDNF in increasing energy expenditure by modulating metabolic rate and temperature. The data indicated that BDNF enhanced energy expenditure suggesting an anorexigenic function [52]. Another finding attested the role of BDNF in thermogenic regulation in lateral hypothalamus [54]. On the contrary, deletion of *Bdnf* gene caused hyperphagia, decreased locomotor activity and impaired thermoregulation [54]. Moreover, it is evident that mutation in the *Bdnf* gene or its receptor (TrkB) leads to obesity in mice [55, 56]. The *Bdnf* gene mutation data is corroborated with hyperphagia and impaired cognitive functions in humans [57–62]. Together, it is suggested that PVH region is critical in energy balance in the brain.

In addition, BDNF plays a key role in energy management in non-neuronal cells. Selective ablation of BDNF in liver cells in mice showed reduction in hyperglycemia and hyperinsulinemia caused by a high fat diet [63]. Compromised BDNF signaling is also linked with obesity and the metabolic syndrome in humans [64]. Furthermore, BDNF administration reduced serum glucose and insulin in obese *db/db* mice or improvement of glucose tolerance compared to their vehicle treated counterparts [65, 66]. The underlying molecular mechanism may be the interaction of BDNF with glucagon like peptide 1 (GLP1). Gotoh et al. showed that administration of BDNF decreased the portal glucagon level and did not show any effect on insulin [67]. It is also observed that the intraportal administration of GLP-1 increases BDNF levels in the pancreas and reduces glucagon secretion [67]. Recent study also suggested a role on pancreatic-islet-expressed TrkB to promote peripheral insulin secretion [68]. In addition to BDNF and TrkB, the pro-BDNF receptor, p75NTR is suggested to play a role in glucose homeostasis and insulin sensitivity. Conditional knockout of p75NTR showed improvements of glucose and insulin tolerance in adipose and skeletal muscle [68, 69]. Regarding signaling context of BDNF and metabolic homeostasis, it is yet to be defined which receptor mediated action is more appropriate. The rational lies that pro-BDNF exclusively binds to p75NTR and appeared to show an opposite effect to BDNF-TrkB activity [70]. It established that a single nucleotide polymorphism (SNP) in pro domain of BDNF (Val66Met) is linked with neuropsychiatric disorders in humans and seemed to function through p75NTR [71]. The SNP (Val66Met) variant indicated increased appetite in mice via p76NTR [72], along with alteration of anxiety and anorexic-related behavior [73, 74]. The data may suggest a unique control of energy balance in food intake and anxiety. Finally, the downstream signaling between pro-BDNF and mature BDNF are quite distinct and may appeared to reflect different outcome in neuronal cells. TrkB promotes MAPK/ERK, PI3K, and PLCg1, pathways, while p75NTR promotes JNK and Rho pathways [36, 41, 75–77].

5. BDNF and psychiatric diseases and disorders

We often use the term disorder and diseases in psychiatric illness. There is a subtle difference exists between them however, they are considered as mental illness. The term disease defines an involuntary response of biological, physiological, or pathological consequences of illness and, the underlying cause can be measured. The disorder defines disturbance of normal physical or mental health status and is a collection of signs and symptoms closely associated with specific disease. In general speaking, we can say that all diseases are disorders but not all disorders are diseases.

BDNF is one of the most widely studied neurotrophin signaling molecules in the brain responsible for neurite growth, maturation of synapses during development, and synaptic plasticity. We have discussed BDNF's biology, receptor alignment for signaling events in the brain. Essentially, BDNF-TrkB signaling, and its intermediate proteins contributed a critical role in different phases of synaptic development and neuroplasticity in the brain [78]. Moreover, BDNF regulates learning and memory process in young and adult humans [79]. Therefore, aberrant expression or imbalance in BDNF level and its cognate TrkB receptor are associated with many psychiatric disorders (diseases) and neurodegenerative diseases. In addition, anomaly of BDNF level and signaling are linked to diverse cardiovascular, metabolic, and inflammatory diseases [80–85]. This section will discuss the contribution of BDNF in brain illness or psychological diseases (disorders) including major depressive disorder (MDD), schizophrenia (SZ), bipolar disorder (BD) and post-traumatic stress disorder (PTSD).

6. BDNF and MDD

BDNF is well studied molecule in MDD. Eisch et al reported that an increase level of BDNF in the ventral tegmental area (VTA)-nucleus accumbens (NAc) region contributed the onset of depression in rats [86]. A following mechanistic study by the same group using viral-mediated mesolimbic dopamine-specific BDNF knockdown determined the pivotal role of BDNF in depression like behavior [87]. Interestingly, reduced BDNF in cornu ammonis (CA3) and dentate gyrus (DG) of the hippocampus and prefrontal cortex (PFC), resulting in depression-like behavior in mice [88]. Furthermore, targeted deletion of BDNF using NSE-tTA x TetOp-Cre line in the VTA area determined that BDNF in the DG was essential for therapeutic intervention as an antidepressant [89]. Similarly, reduced BDNF protein levels were observed in patients with MDD compared with the healthy control [90, 91]. Taken together, these findings suggest that BDNF acts within the VTA-NAc pathway to induce a depression-like phenotype, whereas in the hippocampus and PFC it produces antidepressant-like effects [92]. It is further observed that TrkB, the receptor for BDNF played a role in MDD. Patient with MDD showed elevated level of TrkB compared to the healthy control [93, 94]. However, it is unclear regarding the role of the partners in MDD and may be the focus of future investigation.

Epigenetic modification like DNA methylation is frequently studied in *Bdnf* gene and BDNF exon I and IV promoters. A methylation profile in CpG island of exon I of BDNF promoter showed differential pattern of methylation that can distinguish between major depression vs. and healthy controls and suggested to be a good biomarker for MDD [95]. But exon IV did not show any changes. A similar study reported higher methylation of BDNF exon I promoter in patients with MDD [96]. This study further showed reduced methylation pattern with antidepressants treatment [96]. Interestingly, patient with MDD showed poor treatment response when methylation of CpG site −87 of BDNF exon IV promoter was lacking [97].

An association between BDNF Val66Met polymorphism and MDD is extensively studied. Meta analyses revealed that there is no association between Val66Met polymorphism and MDD (depression) [98–100]. However, few studies have indicated that BDNF Val66Met polymorphism moderated the relationship between stress and depression [100–103].

7. BDNF and Schizophrenia (SZ)

Schizophrenia is a complex heterogenous disease characterized by multiple symptoms such as hallucinations, social avoidance, withdrawal, paranoia, cognitive deficit, and disorganized thought [104]. The role of BDNF in SZ is well studied because BDNF is involved in neurotransmission. In general, BDNF level is reduced in SZ patients [105, 106] and study has shown further that serum BDNF is positively correlated with antipsychotic drug (clozapine) [107]. This is an interesting finding for a therapeutic purpose. However, recent evidence implicated that BDNF mRNA expression remained unchanged in SZ patients compared to healthy control in postmortem brain samples [108].

Reports are emerging regarding epigenetic mechanism in *Bdnf* gene and development of SZ [109]. Epigenetic mechanism encompasses DNA methylation, histone modification, chromatin remodeling and DNA methylation is widely studied in SZ [109, 110]. A significant positive correlation was observed in BDNF gene methylation in patients with SZ compared to healthy controls [111]. Another study showed higher methylation level at BDNF promoter compared to controls [112]. Moreover, a differentially methylated CpGs has been identified in SZ patients of postmortem human brains [113]. Moreover the Val66Met SNP on the *Bdnf* gene has implicated schizophrenia incidence and a recent meta-analysis provided evidence that there was an association between brain volume alterations and variations on the Val66Met SNP in patients of SZ [114–116]. While studies have shown a positive correlation between reduced level of BDNF and SZ episode, but have not evaluated the role of demographic characteristics such as age, gender, race, and education. Therefore, adequate meta-analysis including demographic factors should be added and warranted further investigation.

8. BDNF and bipolar disorder (BD)

Bipolar disorder is a multifactorial psychiatric disorder characterized by mood fluctuation or instability, depressive, manic episode, and euthymic states [117, 118]. BD makes a distinct category in Diagnostic and Statistical Manual of Mental Disorders, 5th edition into BD I, BD II based on severity of manic episodes [119]. The thirst for potential biomarker in BP is emerging and BDNF is extensively studied in this area. In 2005, Laske et al. first reported reduced BDNF level in the serum of manic and major depressed patients compared to healthy control [120]. Since, then several studies have been conducted in BD and majority of the studies suggested a decline level of peripheral BDNF and considered it as a marker [121–125], however, BDNF levels were not different in euthymia when compared to controls [126]. Furthermore, at transcription level, BDNF mRNA showed downregulation in postmortem brains of both manic and depressive subjects [127, 128]. Antipsychotic drugs like mood stabilizers are frequently prescribed for manic or depressive disorder but the study did not show any improvement of BDNF level in four weeks treatment [122]. However, another study of sixteen-week follow-up, using extended-release quetiapine showed increase in BDNF levels, but decreases with time in a manic/mixed episode [129].

A common genetic variation in *Bdnf* gene, the Val66Met, is established as a common platform linked with reduced secretion of BDNF and is associated with many neuropsychiatric disorders and BD is not an exception. Earlier finding suggested an association between BDNF Val66Met polymorphism and BP [130, 131] but recent meta-analyses showed opposite results [132, 133]. Therefore, more data are warranted to determine the role of Val66Met polymorphism in BD.

Epigenetic modulation is well documented in psychiatric disorders and a positive correlation is shown in CpG methylation in BDNF promoter and BD subjects [134–136]. Alterations in DNA methylation patterns in patients with BD have been extensively investigated for the past years, and possibly recognize a potential biomarker [137–139]. It may be the case that DNA methylation alters the differences in BDNF level and contributed in part in BD, so, targeting BDNF methylation could be strategy to treat BD.

9. BDNF and post-traumatic stress disorder

Post-traumatic stress disorder (PTSD) is a debilitating psychiatric disorder characterized by hyperarousal, re-experiencing, negative emotions, increased anxiety, and fearful memories following exposure to severe trauma [119]. The role of BDNF in PTSD is emerging. In 2009, a small human study was conducted in University of Pisa, Italy where they recruited 18 drug naïve PTSD patients (12 women and 6 men) with no psychiatric comorbidity and 18 healthy controls in outpatients' facility. The finding showed reduced level of BDNF in the plasm compared with healthy control [140]. War Veterans have continuously suffered from PTSD and cognitive deficit caused by traumatic brain injury. The possible first combat Veteran study aiming BDNF as a marker in PTSD was investigated in Croatia, 2022. The results revealed a marked reduction in plasma BDNF in Veterans with PTSD and mild cognitive impairment compared with healthy controls [141]. The epigenetic influence in BDNF played a critical role in psychiatric disorders including PTSD, as few studies were conducted to investigate DNA methylation in CpG island and Val66Met polymorphisms. A study was conducted using US military service members deployed in the Middle east for Operation Iraqi Freedom (OIF)/Operation Enduring Freedom (OEF) with PTSD showing a significant association between BDNF Val66 Met genotype and traumatic stress in post deployment [142]. Another study of Vietnam war active service members from South Korea showed an association between higher DNA methylation in BDNF promoter in PTSD subjects suggesting a biomarker of PTSD [143]. Interestingly, another study of Vietnam war Veterans by the Australian or New Zealand Defense Force showed that PTSD was associated with decreased methylation at three BDNF CpG sites [144]. Furthermore, it was observed that BDNF Val66Met was linked with differential *Bdnf* expression in the peripheral tissues [144]. Another study supported the finding that methylation of CpG island (CpG1, CpG 7 and CpG 18) in BDNF promoter was closely related to PTSD and suggested as a biomarker to PTSD [145].

Although studies have shown a positive correlation between BDNF level and Val66Met polymorphism in PTSD, there were reports that showed the opposite effect. There was a report showing no relationship between BDNF Val66Met and PTSD in victims of urban violence [146]. In addition, two case studies (small sample size) failed to establish the association between Val66Met and PTSD [147, 148]. Moreover, an elevated level of BDNF was observed in patients with PTSD suffering from trauma [149]. A meta-analysis showed that BDNF level is increased in PTSD patients compared to healthy subjects [150]. A discrepancy was noted in OEF/OIF Veteran study. Recently, Wu et al. reported for the first time that a higher serum level of BDNF in chronic combat PTSD Veterans independent of symptom severity [151]. These reports contradict previous findings.

Together it appeared that genetic variants of *Bdnf* gene and PTSD did not provide any conclusive relationship. The higher and lower value of BDNF were possibly observed due to heterogenous population or low percentage of homozygous Met alleles. More longitudinal and follow-up studies are necessary to make a definitive conclusion.

10. BDNF-miRNAs-psychiatric disorders

The miRNAs are non-coding RNAs, a new class of epigenetic modulators emerging as an attractive molecule for therapeutic intervention. The miRNAs are small 21–23 nucleotides that have the capability to inhibit mRNA and protein resulting in gene regulation [152]. Literature search showed 2844 articles have been published where miRNAs were associated with psychiatric diseases. Interestingly, BDNF-miRNA axis in psychiatric diseases showed 131 reports indicating therapeutic potential of BDNF. Recent studies indicated that several miRNAs target 3′ UTR of *Bdnf* gene modulated the function associated with psychiatric disorders [153–158].

In rodent model of anxiety disorder and schizophrenia, miR-124a regulated anxiety like behavior by targeting *Bdnf* gene [159] and miR-148b is implicated in regulating *Bdnf* gene in methylazoxymethanol acetate model [160]. In mouse model of PTSD, a set of miRNAs, miR-15a-5p, miR-497a-5p, miR-511-5p and let-7d-5p were shown to be associated with *Bdnf* and *FKBP5*, the two key PTSD-linked genes [157]. Moreover, a prolong stress induced rat PTSD model, miR-142-5p is shown to be upregulated in amygdala with a target gene, Npas4 which was reduced [161]. The inhibition of miR-142-5p appeared to reduce the PTSD symptoms by restoring Npas4 and BDNF level suggesting a crucial link between them. In BD condition, a human cohort study was conducted and revealed an association between miR-206 and BDNF polymorphism [162]. Another study showed a panel of miRNAs, miR-7-5p, miR-221-5p and miR-370-5p that are involved in BD II patients by modulating BDNF level [163].

In summary, the data showed promising direction in miRNA-BDNF-axis modulation in psychiatric disorders. However, a strong clinical correlation regarding miRNA-BDNF needs to be established for the development of new diagnostic and therapeutic application to mitigate the cognitive deficit.

11. Conclusion

BDNF is well studied in major psychiatric disorders or diseases. Modern techniques provided us new insights regarding BDNF's role in psychiatric disease progression and treatment responses. The dysregulation of BDNF/pro-BDNF and its receptors TrkBs resulting in a cascade of neuropathophysiological events leading to the impairment of synaptic plasticity and cognitive deficit. Several lines of evidence support the notion that BDNF is nodal mediator across an array of neuropsychiatric disorders. It is further to make a note that many second-generation antipsychotic drugs showed some promise in providing neuroprotection by enhancing BDNF level, however, a definitive conclusion cannot be made based on few medications. Future investigation including using small molecule compound (mimetics or agonists) for enhancing BDNF synthesis and gene therapy using nanoparticle mediated encapsulation of BDNF, is necessary to extend this efficacy at therapeutic standpoint. Peripheral BDNF level is used as a biomarker in many psychiatric disorders, however, in some cases like MDD, it showed disagreement. This may be due to heterogenous nature and epigenetic modifications that contributed significantly for making a universal conclusion. Nonetheless, it helped to pave the way for better understanding the role of BDNF deep inside human brains. Future studies are warranted to uncover the mechanism of methylation and SNPs of *Bdnf* gene for better therapeutic treatment.

Acknowledgements

This material is the result of work with resources and the use of facilities at the VISN17 Center of Excellence for Research on Returning War Veterans and the Central Texas Veterans Health Care System (CTVHCS). The authors would like to acknowledge Richard W. Seim, Ph.D., Director of the VISN 17 Center of Excellence for Research on Returning War Veterans, (Texas Waco, USA) for his support. The views expressed herein are those of the authors and do not necessarily reflect the official policy or position of the Department of Veterans Affairs or the United States Government.

Funding

The present study was supported by VISN17 Center of Excellence's internal funds to Biomarkers & Genetics Core.

Conflict of interest

The authors state that they do not have any conflict of interest.

Notes/thanks/declaration

None.

Author details

Sudhiranjan Gupta* and Rakeshwar S. Guleria
Biomarkers and Genetics Core, VISN 17 Center of Excellence for Research on Returning War Veterans, Central Texas Veterans Health Care System, Waco, Texas, United States

*Address all correspondence to: sudhiranjan.gupta@va.gov

References

[1] Barde YA, Edgar D, Thoenen H. Purification of a new neurotrophic factor from mammalian brain. The EMBO Journal. 1982;**1**:549-553. DOI: 10.1002/j.1460-2075. 1982.tb01207. x

[2] Zafra F, Hengerer B, Leibrock J, Thoenen H, Lindholm D. Activity dependent regulation of BDNF and NGF mRNAs in the rat hippocampus is mediated by non-NMDA glutamate receptors. The EMBO Journal. 1990;**9**:3545-3550. DOI: 10.1002/j.1460-2075. 1990.tb07564. x

[3] Dugich-Djordjevic MM, Tocco G, Lapchak PA, Pasinetti GM, Najm I, Bundry M, et al. Regionally specific and rapid increases in brain-derived neurotrophic factor messenger RNA in the adult rat brain following seizures induced by systemic administration of kainic acid. Neuroscience. 1992;**47**:303-315. DOI: 10.1016/0306-4522(92)90246-x

[4] Thoenen H. Neurotrophins and neuronal plasticity. Science. 1995;**270**:593-598. DOI: 10.1126/science.270.5236.593

[5] Lindsay RM, Wiegand SJ, Anthony Altar CA, DiStefano P. Neurotrophic factors: From molecule to man. Trends in Neurosciences. 1994;**17**:182-190. DOI: 10.1016/0166-2236(94)90099-x

[6] Lindvall O, Kokaia Z, Bengzon J, Elmer E, Kokaia M. Neurotrophins and brain insults. Trends in Neurosciences. 1994;**17**:490-496. DOI: 10.1016/0166-2236(94)90139-2

[7] Siuciak JA, Anthony Altar CA, Wiegand SJ, Lindsay RM. Antinociceotive effect of brain-derived neurotrophic factor and neurotrophin-3. Brain Research. 1994;**633**:326-330. DOI: 10.1016/0006-8993(94)91556-3

[8] Shirayama Y, Chen AC, Nakagawa S, Russell DS, Duman RS. Brain-derived neurotrophic factor produces antidepressant effects in behavioral models of depression. The Journal of Neuroscience. 2002;**22**:3251-3261. DOI: 10.1523/JNEUROSCI.22-08-03251.2002

[9] Siuciak JA, Lewis DR, Wiegand SJ, Lindsay RM. Antidepressant-like effect of brain-derived neurotrophic factor (BDNF). Pharmacology, Biochemistry, and Behavior. 1997;**56**:131-137. DOI: 10.1016/S0091-3057(96)00169-4

[10] Altar CA. Neurotrophins and depression. Trends in Pharmacological Sciences. 1999;**20**:59-61. DOI: 10.1016/s0165-6147(99)01309-7

[11] Nibuya M, Morinobu S, Duman RS. Regulation of BDNF and trkB mRNA in rat brain by chronic electroconvulsive seizure and antidepressant drug treatments. The Journal of Neuroscience. 1995;**15**:7539-7547. DOI: 10.1523/JNEUROSCI.15-11-07539.1995

[12] Tapia-Arancibia L, Aliaga E, Silhol M, Arancibia S. New insights into brain BDNF function in normal aging and Alzheimer disease. Brain Research Reviews. 2008;**59**:201-220. DOI: 10.1016/j.brainresrev.2008.07.007

[13] Lee J, Duan W, Mattson MP. Evidence that brain-derived neurotrophic factor is required for basal neurogenesis and mediates, in part, the enhancement of neurogenesis by dietary restriction in the hippocampus of adult mice. Journal of Neurochemistry. 2002;**82**:1367-1375. DOI: 10.1046/j.1471-4159.2002.01085. x

[14] Lu B, Nagappan G, Lu Y. BDNF and synaptic plasticity, cognitive function, and dysfunction. In: Lewin G, Carter B, editors. Neurotrophic Factors. Handbook of Experimental Pharmacology. Berlin/Heidelberg, Germany: Springer; 2014. pp. 223-250. DOI: 10.1007/978-3-642-45106-5_9

[15] Zuccato C, Cattaneo E. Brain-derived neurotrophic factor in neurodegenerative diseases. Nature Reviews. Neurology. 2009;**5**:311-322. DOI: 10.1038/nrneurol.2009.54

[16] Autry AE, Monteggia LM. Brain-derived neurotrophic factor and neuropsychiatric disorders. Pharmacological Reviews. 2012;**64**:238-258. DOI: 10.1124/pr.111.005108

[17] Levi-Montalcini R, Hamburger V. Selective growth stimulating effects of mouse sarcoma on the sensory and sympathetic nervous system of chick embryo. The Journal of Experimental Zoology. 1951;**116**:321-361. DOI: 10.1002/jez.1401160206

[18] Leibrock J, Lottspeich F, Hohn A, Hofer M, Hengerer B, Masiakowski P, et al. Molecular cloning and expression of brain-derived neurotrophic factor. Nature. 1989;**341**:149-152. DOI: 10.1038/341149a0

[19] Radziejewski C, Robinson RC, DeStefano PS, Taylor JW. Dimeric structure and conformational stability of brain-derived neurotrophic factor and neurotrophin-3. Biochemistry. 1992;**31**:4431-4436. DOI: 10.1021/bi00133a007

[20] Mowla SJ, Farhadi HF, Pareek S, Atwal JK, Morris SJ, Seidah NG, et al. Biosynthesis and post-translational processing of the precursor to brain-derived neurotrophic factor. The Journal of Biological Chemistry. 2001;**276**(16):12660-12666. DOI: 10.1074/jbc.M008104200

[21] Foltran RB, Diaz SL. BDNF isoforms: A round trip ticket between neurogenesis and serotonin? Journal of Neurochemistry. 2016;**138**(2):204-221. DOI: 10.1111/jnc.13658

[22] Dieni S, Matsumoto T, Dekkers M, Rauskolb S, Ionescu MS, Deogracias R, et al. Barde YA BDNF and its pro-peptide are stored in presynaptic dense core vesicles in brain neurons. The Journal of Cell Biology. 2012;**196**(6):775-788. DOI: 10.1083/jcb.201201038

[23] Yang J, Harte-Hargrove LC, Siao CJ, Marinic T, Clarke R, Ma Q, et al. proBDNF negatively regulates neuronal remodeling, synaptic transmission, and synaptic plasticity in hippocampus. Cell Reports. 2014;7(3):796-806. DOI: 10.1016/j.celrep.2014.03.040

[24] Lessmann V, Brigadski T. Mechanisms, locations, and kinetics of synaptic BDNF secretion: An update. Neuroscience Research. 2009;**65**:11-22. DOI: 10.1016/j.neures.2009.06.004

[25] Lee J, Seroogy KB, Mattson MP. Dietary restriction enhances neurotrophins expression and neurogenesis in the hippocampus of adult mice. Journal of Neurochemistry. 2002;**80**:539-547. DOI: 10.1046/j.0022-3042.2001.00747. x

[26] Brigadski T, Leßmann V. The physiology of regulated BDNF release. Cell and Tissue Research. 2020;**382**(1):15-45. DOI: 10.1007/s00441-020-03253-2

[27] Pruunsild P, Kazantseva A, Aid T, Palm K, Timmusk T. Dissecting the human BDNF locus: Bidirectional transcription, complex splicing, and multiple promoters. Genomics.

2007;**90**:397-406. DOI: 10.1016/j.ygeno.2007.05.004

[28] Adachi N, Numakawa T, Richards M, Nakajima S, Kunugi H. New insight in expression, transport, and secretion of brain-derived neurotrophic factor: Implications in brain-related diseases. World Journal of Biological Chemistry. 2014;**5**:409-428. DOI: 10.4331/wjbc.v5.i4.409

[29] Bergami M, Rimondini R, Santi S, Blum R, Gotz M, Canossa M. Deletion of TrkB in adult progenitors alters newborn neuron integration into hippocampal circuits and increases anxiety-like behavior. Proceedings of the National Academy of Sciences of the United States of America. 2008;**105**(40):15570-15575. DOI: 10.1073/pnas.0803702105

[30] Vilar M, Mira H. Regulation of neurogenesis by neurotrophins during adulthood: Expected and unexpected roles. Frontiers in Neuroscience. 2016;**10**:26. DOI: 10.3389/fnins.2016.00026

[31] Lu Y, Christian K, Lu B. BDNF: A key regulator for protein-synthesis dependent LTP and long-term memory? Neurobiology of Learning and Memory. 2008;**89**(3):312-323. DOI: 10.1016/j.nlm.2007.08.018

[32] Lyons WE, Mamounas LA, Ricaurte GA, Coppola V, Reid SW, Bora SH, et al. Brain-derived neurotrophic factor-deficient mice develop aggressiveness and hyperphagia in conjunction with brain serotonergic abnormalities. Proceedings of the National Academy of Sciences of the United States of America. 1999;**96**:15236-15244. DOI: 10.1073/pnas.96.26.15239

[33] Molteni R, Barnard RJ, Ying Z, Roberts CK, Gomez-Pinilla F. A high-fat, refined sugar diet reduces hippocampal brain-derived neurotrophic factor, neuronal plasticity, and learning. Neuroscience. 2002;**112**:803-814. DOI: 10.1016/s0306-4522(02)00123-9

[34] Kaplan DR, Hempstead BL, Martin-Zanca D, Chao MV, Parada LF. The trk proto-oncogene product: A signal transducing receptor for nerve growth factor. Science. 1991;**252**:554-558. DOI: 10.1126/science.1850549

[35] Barbacid M. The trk family of neurotrophin receptors. Journal of Neurobiology. 1994;**25**:1386-1403. DOI: 10.1126/science.1850549

[36] Yoshii A, Constantine-Paton M. Postsynaptic BDNF-TrkB signaling in synapse maturation, plasticity, and disease. Developmental Neurobiology. 2010;**70**:304-322. DOI: 10.1002/dneu.20765

[37] Chao M, Hempstead B. p75 and Trk: A two-receptor system. Trends in Neurosciences. 1995;**18**(7):321-326

[38] Mohammadi A, Amooeian VG, Rashidi E. Dysfunction in brain-derived neurotrophic factor signaling pathway and susceptibility to schizophrenia. Parkinson's and Alzheimer's diseases. Current Gene Therapy. 2018;**18**:45-63. DOI: 10.2174/1566523218666180302163029

[39] Suzuki S, Numakawa T, Shimazu K, Koshimizu H, Hara T, Hatanaka H, et al. BDNF-induced recruitment of TrkB receptor into neuronal lipid rafts: Roles in synaptic modulation. The Journal of Cell Biology. 2004;**167**(6):1205-1215. DOI: 10.1083/jcb.200404106

[40] Gonzalez A, Moya-Alvarado G, Gonzalez-Billaut C, Bronfman FC. Cellular and molecular mechanisms regulating neuronal growth by brain-derived neurotrophic factor

(BDNF). Cytoskeleton (Hoboken). 2016;**73**(10):612-616. DOI: 10.1002/cm.21312

[41] Huang EJ, Reichardt LF. Trk receptors: Roles in neuronal signal transduction. Annual Review of Biochemistry. 2003;**72**:609-642. DOI: 10.1146/annurev.biochem.72.121801.161629

[42] Baydyuk M, Xu B. BDNF signaling and survival of striatal neurons. Frontiers in Cellular Neuroscience. 2014;**8**:254. DOI: 10.3389/fncel.2014.00254

[43] Zuccato C, Cattaneo. Role of brain-derived neurotrophic factor in Huntington disease. Progress in Neurobiology. 2007;**81**:294-311. DOI: 10.1016/j.pneurobio.2007.01.003

[44] Zhang SJ, Zou M, Lu L, Lau D, Ditzel DAW, Delucinge-Vivier C, et al. Nuclear calcium signaling controls expression of a large gene pool: Identification of a gene program for acquired neuroprotection induced by synaptic activity. PLoS Genetics. 2009;**5**(8):e1000604. DOI: 10.1371/journal.pgen.1000604

[45] Reichardt LF. Neurotrophin-regulated signalling pathways. Philosophical Transactions of the Royal Society of London. Series B, Biological Sciences. 2006;**361**(1473):1545-1564. DOI: 10.1098/rstb.2006.1894

[46] Minichiello L. TrkB signalling pathways in LTP and learning. Nature Reviews. Neuroscience. 2009;**10**(12):850-860. DOI: 10.1038/nrn2738

[47] Kwon M, Fernandez JR, Zegarek GF, Lo SB, Firestein BL. BDNF-promoted increases in proximal dendrites occur via CREB dependent transcriptional regulation of cypin. The Journal of Neuroscience. 2011;**31**(26):9735-9745. DOI: 10.1523/JNEUROSCI.6785-10.2011

[48] Chao MV. Neurotrophin signaling in health and disease. Clinical Science (London, England). 2006;**110**:167-173. DOI: 10.1042/CS20050163

[49] Noble EE, Billington CJ, Kotz CM, Wang C. The lighter side of BDNF. American Journal of Physiology. Regulatory, Integrative and Comparative Physiology. 2011;**300**:R1053-R1069. DOI: 10.1152/ajpregu.00776.2010

[50] Podyma B, Parekh K, Güler AD, Deppmann CD. Metabolic homeostasis via BDNF and its receptors. Trends in Endocrinology and Metabolism. 2021;**32**:488-499. DOI: 10.1016/j.tem.2021.04.005

[51] Wang C, Bomberg E, Billington C, Levine A, Kotz CM. Brain-derived neurotrophic factor in the hypothalamic paraventricular nucleus increases energy expenditure by elevating metabolic rate. American Journal of Physiology. Regulatory, Integrative and Comparative Physiology. 2007;**293**:R992-R1002. DOI: 10.1152/ajpregu.00516.2006

[52] Tsuchida A, Nonomura T, Ono-Kishino M, Nakagawa T, Taiji M, Noguchi H. Acute effects of brain-derived neurotrophic factor on energy expenditure in obese diabetic mice. International Journal of Obesity and Related Metabolic Disorders. 2001;**25**:1286-1293. DOI: 10.1038/sj.ijo.0801678

[53] Cao L, Choi EY, Liu X, Martin A, Wang C, Xu X, et al. White to brown fat phenotypic switch induced by genetic and environmental activation of a hypothalamic-adipocyte axis. Cell Metabolism. 2011;**14**:324-338. DOI: 10.1016/j.cmet.2011.06.020

[54] You H, Chu P, Guo W, Lu B. A subpopulation of Bdnf-e1-expressing glutamatergic neurons in the lateral hypothalamus critical for thermogenesis control. Molecular Metabolism. 2020;**31**:109-123. DOI: 10.1016/j.molmet.2019.11.013

[55] An JJ, Liao G-Y, Kinney CE, Sahibzada N, Xu B. Discrete BDNF neurons in the paraventricular hypothalamus control feeding and energy expenditure. Cell Metabolism. 2015;**22**:175-188. DOI: 10.1016/j.cmet.2015.05.008

[56] Rios M, Fan G, Fekete C, Kelly J, Bates B, Kuehn R, et al. Conditional deletion of brain-derived neurotrophic factor in the postnatal brain leads to obesity and hyperactivity. Molecular Endocrinology. 2001;**15**:1748-1757. DOI: 10.1210/mend.15.10.0706

[57] Liao GY, An JJ, Gharami K, Waterhouse EG, Vanevski F, Jones KR, et al. Dendritically targeted Bdnf mRNA is essential for energy balance and response to leptin. Nature Medicine. 2012;**18**:564-571. DOI: 10.1038/nm.2687

[58] Gray J, Yeo GSH, Cox JJ, Morton J, Adlam A-LR, Keogh JM, et al. Hyperphagia, severe obesity, impaired cognitive function, and hyperactivity associated with functional loss of one copy of the brain-derived neurotrophic factor (BDNF) gene. Diabetes. 2006;**55**:3366-3371. DOI: 10.2337/db06-0550

[59] Mou Z, Hyde TM, Lipska BK, Martinowich K, Wei P, Ong C-J, et al. Human obesity associated with an intronic SNP in the brain-derived neurotrophic factor locus. Cell Reports. 2015;**13**:1073-1080. DOI: 10.1016/j.celrep.2015.09.065

[60] Han JC, Liu QR, Jones M, Levinn RL, Menzie CM, Jefferson-George KS, et al. Brain-derived neurotrophic factor and obesity in the WAGR syndrome. The New England Journal of Medicine. 2008;**359**:918-927. DOI: 10.1056/NEJMoa0801119

[61] Xu B, Xie X. Neurotrophic factor control of satiety and body weight. Nature Reviews. Neuroscience. 2016;**17**:282-292. DOI: 10.1038/nrn.2016.24

[62] Rios M. Neurotrophins and the regulation of energy balance and body weight. Handbook of Experimental Pharmacology. 2014;**220**:283-307. DOI: 10.1007/978-3-642-45106-5_11

[63] Teillon S, Calderon GA, Rios M. Diminished diet-induced hyperglycemia and dyslipidemia and enhanced expression of PPARalpha and FGF21 in mice with hepatic ablation of brain-derived neurotropic factor. The Journal of Endocrinology. 2010;**205**:37-47. DOI: 10.1677/JOE-09-0405. doi: 10.1677/JOE-09-0405

[64] Yang AC, Chen TJ, Tsi SJ, Hong C, Kuo CH, Yag CH, et al. BDNF Val66Met polymorphism alters sympathovagal balance in healthy subjects. American Journal of Medical Genetics. Part B, Neuropsychiatric Genetics. 2010;**153B**:1024-1030. DOI: 10.1002/ajmg.b.31069

[65] Nonomura T, Tsuchida A, Ono-Kishino M, Nakagawa T, Taiji M, Noguchi H. Brain-derived neurotrophic factor regulates energy expenditure through the central nervous system in obese diabetic mice. International Journal of Experimental Diabetes Research. 2001;**2**:201-209. DOI: 10.1155/edr.2001.201

[66] Nakagawa T, Ogawa Y, Ebihara K, Yamanaka M, Tsuchida A, Taiji M, et al. Anti-obesity and anti-diabetic effects

of brain-derived neurotrophic factor in rodent models of leptin resistance. International Journal of Obesity and Related Metabolic Disorders. 2003;**27**:557-565. DOI: 10.1038/sj.ijo.0802265

[67] Gotoh K, Masaki T, Chiba S, Ando H, Fujiwara K, Shimasaki T, et al. Hypothalamic brain-derived neurotrophic factor regulates glucagon secretion mediated by pancreatic efferent nerves. Journal of Neuroendocrinology. 2013;**25**:302-311. DOI: 10.1111/jne.12003

[68] Fulgenzi G, Hong Z, Tomassoni-Ardori F, Barella LF, Becker J, Barrick C, et al. Novel metabolic role for BDNF in pancreatic β-cell insulin secretion. Nature Communications. 2020;**11**:1950. DOI: 10.1038/s41467-020-15833-5

[69] Baeza-Raja B, Li P, Le Moan N, Sachs BD, Schachtrup C, Davalos D, et al. p75 neurotrophin receptor regulates glucose homeostasis and insulin sensitivity. Proceedings of the National Academy of Sciences of the United States of America. 2012;**109**:5838-5843. DOI: 10.1073/pnas.1103638109

[70] Baeza-Raja B, Sachs BD, Li P, Christian F, Vagena E, Davalos D, et al. p75 neurotrophin receptor regulates energy balance in obesity. Cell Reports. 2016;**14**:255-268. DOI: 10.1016/j.celrep.2015.12.028

[71] Teng HK, Teng KK, Lee R, Wright S, Tevar S, Almeida RD, et al. ProBDNF induces neuronal apoptosis via activation of a receptor complex of p75NTR and sortilin. The Journal of Neuroscience. 2005;**25**:5455-5463. DOI: 10.1523/JNEUROSCI.5123-04.2005

[72] Anastasia A, Deinhardt K, Chao MV, Will NE, Irmady K, Lee FS, et al. Val66Met polymorphism of BDNF alters prodomain structure to induce neuronal growth cone retraction. Nature Communications. 2013;**4**:2490. DOI: 10.1038/ncomms3490

[73] Ieraci A, Barbieri SS, Macchi C, Amadio P, Sandrini L, Magni P, et al. BDNF Val66Met polymorphism alters food intake and hypothalamic BDNF expression in mice. Journal of Cellular Physiology. 2020;**235**:9667-9675. DOI: 10.1002/jcp.29778

[74] Chen Z-Y, Jing D, Bath KG, Ieraci A, Khan T, Siao CJ, et al. Genetic variant BDNF (Val66Met) polymorphism alters anxiety-related behavior. Science. 2006;**314**:140-143. DOI: 10.1126/science.1129663

[75] Madra M, Zeltser LM. BDNF-Val66Met variant and adolescent stress interact to promote susceptibility to anorexic behavior in mice. Translational Psychiatry. 2016;**6**:e776. DOI: 10.1038/tp.2016.35

[76] Lin Z, Tann JY, Goh ETH, Kelly C, Lim KB, Gao JF, et al. Structural basis of death domain signaling in the p75 neurotrophin receptor. eLife. 2015;**4**:e11692. DOI: 10.7554/eLife.11692

[77] Harrington AW, Kim JY, Yoon SO. Activation of Rac GTPase by p75 is necessary for c-Jun N-terminal kinase-mediated apoptosis. The Journal of Neuroscience. 2002;**22**:156-166. DOI: 10.1523/JNEUROSCI.22-01-00156.2002

[78] Edelmann E, Lessmann V, Brigadski T. Pre- and postsynaptic twists in BDNF secretion and action in synaptic plasticity. Neuropharmacology. 2014;**76**(Pt C):610-627. DOI: 10.1016/j.neuropharm.2013.05.043

[79] Boschen KE, Klintsova AY. Neurotrophins in the brain: Interaction

with alcohol exposure during development. Vitamins and Hormones. 2017;**104**:197-242. DOI: 10.1016/bs.vh.2016.10.008

[80] Hang PZ, Zhu H, Li PF, Liu J, Ge FQ, Zhao J, et al. The emerging role of BDNF/TrkB signaling in cardiovascular diseases. Life (Basel). 2021;**11**(1):70. DOI: 10.3390/life11010070

[81] Kermani P, Hempstead B. BDNF actions in the cardiovascular system: Roles in development, adulthood and response to injury. Frontiers in Physiology. 2019;**26**(10):455. DOI: 10.3389/fphys.2019.00455

[82] Eyileten C, Kaplon-Cieslicka A, Mirowska-Guzel D, Malek L, Postula M. Antidiabetic effect of brain-derived neurotrophic factor and its association with inflammation in type 2 diabetes mellitus. Journal Diabetes Research. 2017;**2017**:2823671

[83] Rozanska O, Uruska A, Zozulinska-Ziolkiewicz D. Brain derived neurotrophic factor and diabetes. International Journal of Molecular Sciences. 2020;**21**(3):841. DOI: 10.3390/ijms21030841

[84] Prakash YS, Martin RJ. Brain-derived neurotrophic factor in the airways. Pharmacology & Therapeutics. 2014;**143**(1):74-86. DOI: 10.1016/j.pharmthera.2014.02.006

[85] Giacobbo BL, Doorduin J, Klein HC, Dierckx RAJO, Bromberg E. de Vries EFJ brain-derived neurotrophic factor in brain disorders: Focus on neuroinflammation. Molecular Neurobiology. 2019;**56**(5):3295-3312. DOI: 10.1007/s12035-018-1283-6

[86] Eisch AJ, Bolaños CA, de Wit J, Simonak RD, Pudiak CM, Barrot M, et al. Brain-derived neurotrophic factor in the ventral midbrain-nucleus accumbens pathway: A role in depression. Biological Psychiatry. 2003;**54**(10):994-1005. DOI: 10.1016/j.biopsych.2003.08.003

[87] Berton O, McClung CA, Dileone RJ, Krishnan V, Renthal W, Russo SJ, et al. Essential role of BDNF in the mesolimbic dopamine pathway in social defeat stress. Science. 2006;**311**(5762):864-868. DOI: 10.1126/science.1120972

[88] Zhang JC, Wu J, Fujita Y, Yao W, Ren Q, Yang C, et al. Antidepressant effects of TrkB ligands on depression-like behavior and dendritic changes in mice after inflammation. The International Journal of Neuropsychopharmacology. 2015;**18**(4):pyu077. DOI: 10.1093/ijnp/pyu077

[89] Adachi M, Barrot M, Autry AE, Theobald D, Monteggia LM. Selective loss of brain-derived neurotrophic factor in the dentate gyrus attenuates antidepressant efficacy. Biological Psychiatry. 2008;**63**(7):642-649. DOI: 10.1016/j.biopsych.2007.09.019

[90] Hsieh MT, Lin CC, Lee CT, Huang TL. Abnormal brain-derived neurotrophic factor exon IX promoter methylation, protein, and mRNA levels in patients with major depressive disorder. Journal of Clinical Medicine. 2019;**8**:568. DOI: 10.3390/jcm8050568

[91] Dwivedi Y, Rizavi HS, Conley RR, Roberts RC, Tamminga CA, Pandey GN. Altered gene expression of brain-derived neurotrophic factor and receptor tyrosine kinase B in postmortem brain of suicide subjects. Archives of General Psychiatry. 2003;**60**(8):804-815. DOI: 10.1001/archpsyc.60.8.804

[92] Hashimoto K, Shimizu E, Iyo M. Critical role of brain-derived neurotrophic factor in mood disorders. Brain Research. Brain Research Reviews.

2004;**45**(2):104-114. DOI: 10.1016/j.brainresrev.2004.02.003

[93] Hung YY, Lin CJ, Huang TL. Higher serum tropomyosin-related kinase B protein level in major depression. Progress in Neuro-Psychopharmacology and Biological Psychiatry. 2010;**34**:610-612. DOI: 10.1016/j.pnpbp.2010.02.021

[94] Tsai SJ. Down-regulation of the Trk-B signal pathway: The possible pathogenesis of major depression. Medical Hypotheses. 2004;**62**:215-218. DOI: 10.1016/S0306-9877(03)00299-8

[95] Fuchikami M, Morinobu S, Segawa M, Okamoto Y, Yamawaki S, Ozaki N. DNA methylation profiles of the brain-derived neurotrophic factor (BDNF) gene as a potent diagnostic biomarker in major depression. PLoS One. 2011:**6**(8):e23881 DOI: 10.1371/journal.pone.0023881

[96] D'Addario C, Dell'Osso B, Galimberti D, Palazzo MC, Benatti B, Di Francesco A. Epigenetic modulation of BDNF gene in patients with major depressive disorder. Biological Psychiatry. 2013;**73**:e6-e7. DOI: 10.1016/j.biopsych.2012.07.009

[97] Tadic A, Muller-Engling L, Schlicht KF, Kotsiari A, Dreimuller N, Kleimann A. Methylation of the promoter of brain-derived neurotrophic factor exon IV and antidepressant response in major depression. Molecular Psychiatry. 2014;**19**:281-283. DOI: 10.1038/mp.2013.58

[98] Verhagen M, van der Meij A, van Deurzen PA, Janzing JG, Arias-Vasquez A, Buitelaar JK. Meta-analysis of the BDNF Val66Met polymorphism in major depressive disorder: Effects of gender and ethnicity. Molecular Psychiatry. 2010;**15**:260-271. DOI: 10.1038/mp.2008.109

[99] Gyekis JP, Yu W, Dong S, Wang H, Qian J, Kota P. No association of genetic variants in BDNF with major depression: A meta- and gene-based analysis. American Journal of Medical Genetics. Part B, Neuropsychiatric Genetics: The Official Publication of the International Society of Psychiatric Genetics. 2013;**16**(2B):61-70. DOI: 10.1002/ajmg.b.32122

[100] Schroter K, Brum M, Brunkhorst-Kanaan N, Tole F, Ziegler C, Domschke K. Longitudinal multi-level biomarker analysis of BDNF in major depression and bipolar disorder. European Archives of Psychiatry and Clinical Neuroscience. 2020;**270**:169-181. DOI: 10.1007/s00406-019-01007-y

[101] Hosang GM, Shiles C, Tansey KE, McGuffin P, Uher R. Interaction between stress and the BDNF Val66Met polymorphism in depression: A systematic review and meta-analysis. BMC Medicine. 2014;**12**:7. DOI: 10.1186/1741-7015-12-7

[102] Januar V, Ancelin ML, Ritchie K, Saffery R, Ryan J. BDNF promoter methylation and genetic variation in late-life depression. Translational Psychiatry. 2015;**5**:e619. DOI: 10.1038/tp.2015.114

[103] Badamasi IM, Lye MS, Ibrahim N, Stanslas J. Genetic endophenotypes for insomnia of major depressive disorder and treatment-induced insomnia. Journal of Neural Transmission. 2019;**126**:711-722. DOI: 10.1007/s00702-019-02014-y

[104] Fujimaki K, Takahashi T, Morinobu S. Association of typical versus atypical antipsychotics with symptoms and quality of life in schizophrenia. PLoS One. 2012;7(5):e37087. DOI: 10.1371/journal.pone.0037087

[105] Yang Y, Liu Y, Wang G, Hei G, Wang X, Li RR, et al. Brain-derived

neurotrophic factor is associated with cognitive impairments in first-episode and chronic schizophrenia. Psychiatry Research. 2019;**273**:5281536. DOI: 10.1016/j.psychres.2019.01.051

[106] Favalli G, Li J, Belmonte-De-Abreu P, Wong AH, Daskalakis ZJ. The role of BDNF in the pathophysiology and treatment of schizophrenia. Journal of Psychiatric Research. 2012;**46**:1-11. DOI: 10.1016/j.jpsychires.2011.09.022

[107] Grillo RW, Ottoni GL, Leke R, Souza DO, Portela LV, Lara DR. Reduced serum BDNF levels in schizophrenic patients on clozapine or typical antipsychotics. Journal of Psychiatric Research. 2007;**41**:31-35. DOI: 10.1016/j.jpsychires.2006.01.005

[108] Cheah SY, McLeay R, Wockner LF, Lawford BR, Young RM, Morris CP, et al. Expression and methylation of BDNF in the human brain in schizophrenia. The World Journal of Biological Psychiatry: The Official Journal of the World Federation of Societies of Biological Psychiatry. 2007;**18**(5):392-400. DOI: 10.1080/15622975.2016.1245443

[109] Punzi G, Bharadwaj R, Ursini G. Neuroepigenetics of schizophrenia. Progress in Molecular Biology and Translational Science. 2018;**158**:195-226. DOI: 10.1016/bs.pmbts.2018.04.010

[110] Ibi D, Gonzalez-Maeso J. Epigenetic signaling in schizophrenia. Cellular Signalling. 2015;**27**(10):2131-2136. DOI: 10.1016/j.cellsig.2015.06.003

[111] Kordi-Tamandani DM, Sahranavard R, Torkamanzehi A. DNA methylation and expression profiles of the brain-derived neurotrophic factor (BDNF) and dopamine transporter (DAT1) genes in patients with schizophrenia. Molecular Biology Reports. 2012;**39**:10889-10893. DOI: 10.1007/s11033-012-1986-0

[112] Ikegame T, Bundo M, Murata Y, Kasai K, Kato T, Iwamoto K. DNA methylation of the BDNF gene and its relevance to psychiatric disorders. Journal of Human Genetics. 2013;**58**:434-438. DOI: 10.1038/jhg.2013.65

[113] Jaffe AE, Gao Y, Deep-Soboslay A, Tao R, Hyde TM, Weinberger DR, et al. Mapping DNA methylation across development, genotype and schizophrenia in the human frontal cortex. Nature Neuroscience. 2016;**19**(1):40-47. DOI: 10.1038/nn.4181

[114] Ahmed AO, Kramer S, Hofman N, Flynn J, Hansen M, Martin V, et al. A meta-analysis of brain-derived neurotrophic factor effects on brain volume in schizophrenia: Genotype and serum levels. Neuropsychobiology. 2021;**80**:411-424. DOI: 10.1159/000514126

[115] Zhao X, Huang Y, Chen K, Li D, Han C, Kan Q. The brain-derived neurotrophic factor Val66Met polymorphism is not associated with schizophrenia: An updated meta-analysis of 11,480 schizophrenia cases and 13,490 controls. Psychiatry Research. 2015;**225**:217-220. DOI: 10.1016/j.psychres.2014.11.015

[116] Notaras M, Hill R, van den Buuse M. A role for the BDNF gene Val66Met polymorphism in schizophrenia? A comprehensive review. Neuroscience & Biobehavioral Reviews. 2015;**51**:15-30. DOI: 10.1016/j.neubiorev.2014.12.016

[117] Merinkangas KR, Low NC. The epidemiology of mood disorders. Current Psychiatry Reports. 2004;**6**:411-421. DOI: 10.1007/s11920-004-0004-1

[118] Rihmer Z, Kiss K. Bipolar disorders and suicidal behavior. Bipolar Disorders.

2002;**4**(Suppl 1):21-25. DOI: 10.1034/j.1399-5618.4. s1.3.x

[119] American Psychiatric Association. Diagnostic and Statistical Manual of Mental Disorders, fifth ed., 2013; DSM-5. Arlington, VA: American Psychiatric Association;

[120] Laske C, Stransky E, Eschweiler GW, Klein R, Wittorf A, Leyhe T, et al. Increased BDNF serum concentration in fibromyalgia with or without depression or antidepressants. Journal of Psychiatric Research. 2007;**41**(7):600-605. DOI: 10.1016/j.jpsychires.2006.02.007

[121] Fernandes BS, Molendijk ML, |Köhler CA, Soares JC, Leite CM, Machado-Vieira R, et al. Peripheral brain-derived neurotrophic factor (BDNF) as a biomarker in bipolar disorder: A meta-analysis of 52 studies. BMC Medicine. 2015;**13**:289. DOI: 10.1186/s12916-015-0529-7

[122] Lin CC, Lee CT, Lo YT, Huang TL. Brain-derived neurotrophic factor protein and mRNA levels in patients with bipolar mania - a preliminary study. Biomedical Journal. 2016;**39**:272-276. DOI: 10.1016/j.bj.2016.08.001

[123] Chiou YJ, Huang TL. Brain-derived neurotrophic factor (BDNF) and bipolar disorder. Psychiatry Research. 2019;**274**:395-399. DOI: 10.1016/j.psychres.2019.02.051

[124] Aas M, Dieset I, Morch R, Steen NE, Hope S, Reponen EJ. Reduced brain-derived neurotrophic factor is associated with childhood trauma experiences and number of depressive episodes in severe mental disorders. Schizophrenia Research. 2019;**205**:45-50. DOI: 10.1016/j.schres.2018.08.007

[125] Mora E, Portella MJ, Pinol-Ripoll G, Lopez R, Cuadras D, Forcada I. High BDNF serum levels are associated to good cognitive functioning in bipolar disorder. European Psychiatry. 2019;**60**:97-107. DOI: 10.1016/j.eurpsy.2019.02.006

[126] Fernandes BS, Gama CS, Cereser KM, Yatham LN, Fries GR, Colpo G. Brain-derived neurotrophic factor as a state-marker of mood episodes in bipolar disorders: A systematic review and meta-regression analysis. Journal of Psychiatric Research. 2011;**45**:995-1004. DOI: 10.1016/j.jpsychires.2011.03.002

[127] Kim HW, Rapoport SI, Rao JS. Altered expression of apoptotic factors and synaptic markers in postmortem brain from bipolar disorder patients. Neurobiology of Disease. 2010;**37**:596-603. DOI: 10.1016/j.nbd.2009.11.010

[128] Lin CC, Hung YY, Huang TL. Associations between DNA methylation of promoter exon IX, serum protein and mRNA levels of brain-derived neurotrophic factor in patients with bipolar mania. Neuropsychiatry. 2018;**8**:224-231

[129] Grande I, Kapczinski F, Stertz L, Colpo GD, Kunz M, Cereser KM. Peripheral brain-derived neurotrophic factor changes along treatment with extended-release quetiapine during acute mood episodes: An open-label trial in drug-free patients with bipolar disorder. Journal of Psychiatric Research. 2012;**46**:1511-1514. DOI: 10.1016/j.jpsychires.2012.08.017

[130] Neves-Pereira M, Mundo E, Muglia P, King N, Macciardi F, Kennedy JL. The brain-derived neurotrophic factor gene confers susceptibility to bipolar disorder: Evidence from a family-based association study. American Journal of Human Genetics. 2002;**71**:651-655. DOI: 10.1016/j.jpsychires.2012.08.017

[131] Li M, Chang H, Xiao X. BDNF Val66Met polymorphism and bipolar disorder in European populations: A risk association in case-control, family-based and GWAS studies. Neuroscience and Biobehavioral Reviews. 2016;**68**:218-233. DOI: 10.1016/j.neubiorev.2016.05.031

[132] Wang Z, Li Z, Gao K, Fang Y. Association between brain-derived neurotrophic factor genetic polymorphism Val66Met and susceptibility to bipolar disorder: A meta-analysis. BMC Psychiatry. 2014;**14**:366. DOI: 10.1186/s12888-014-0366-9

[133] Harrisberger F, Smieskova R, Schmidt A, Lenz C, Walter A, Wittfeld K, et al. BDNF Val66Met polymorphism and hippocampal volume in neuropsychiatric disorders: A systematic review and meta-analysis. Neuroscience and Biobehavioral Reviews. 2015;**55**:107-118. DOI: 10.1016/j.neubiorev.2015.04.017

[134] Mill J, Tang T, Kaminsky Z, Khare T, Yazdanpanah S, Bouchard L. Epigenomic profiling reveals DNA-methylation changes associated with major psychosis. American Journal of Human Genetics. 2008;**82**:696-711. DOI: 10.1016/j.ajhg.2008.01.008

[135] Strauss JS, Khare T, De Luca V, Jeremian R, Kennedy JL, Vincent JB. Quantitative leukocyte BDNF promoter methylation analysis in bipolar disorder. International Journal of Bipolar Disorders. 2013;**1**:28. DOI: 10.1186/2194-7511-1-28

[136] D'Addario C, Dell'Osso B, Palazzo MC, Benatti B, Lietti L, Cattaneo E. Selective DNA methylation of BDNF promoter in bipolar disorder: Differences among patients with BDI and BDII. Neuropsychopharmacology, The Official Publication of the American College of Neuropsychopharmacology. 2012;**37**:1647-1655. DOI: 10.1038/npp.2012.10

[137] Stenz L, Zewdie S, Laforge-Escarra T, Prados J, La Harpe R, Dayer A, et al. BDNF promoter I methylation correlates between postmortem human peripheral and brain tissues. Neuroscience Research. 2015;**91**:1-7. DOI: 10.1016/j.neures.2014.10.003

[138] Carlberg L, Scheibelreiter J, Hassler MR, Schloegelhofer M, Schmoeger M, Ludwig B, et al. Brain-derived neurotrophic factor (BDNF)-epigenetic regulation in unipolar and bipolar affective disorder. Journal of Affective Disorders. 2014;**168**:399-406. DOI: 10.1016/j.jad.2014.07.022

[139] Rao JS, Keleshian VL, Klein S, Rapoport SI. Epigenetic modifications in frontal cortex from Alzheimer's disease and bipolar disorder patients. Translational Psychiatry. 2012a;**2**:e132. DOI: 10.1038/tp.2012.55

[140] Dell'osso L, Carmassi C, Del Debbio A, Dell'osso MC, Bianchi C, da Pozzo E, et al. Brain-derived neurotrophic factor plasma levels in patients suffering from post-traumatic stress disorder. Progress in Neuro-Psychopharmacology & Biological Psychiatry. 2009;**33**:899-902. DOI: 10.1016/j.pnpbp.2009.04.018

[141] Domitrovic Spudic S, Nikolac Perkovic M, Uzun S, Nedic Erjavec G, Kozumplik O, Svob Strac D, et al. Reduced plasma BDNF concentration and cognitive decline in veterans with PTSD. Psychiatry Research. 2022;**316**:114772. DOI: 10.1016/j.psychres.2022.114772

[142] Dretsch MN, Williams K, Emmerich T, Crynen G, Ait-Ghezala G, Chaytow H, et al.

Brain derived neurotropic factor polymorphisms, traumatic stress, mild traumatic brain injury, and combat exposure contribute to postdeployment traumatic stress. Brain and Behavior: A Cognitive Neuroscience Perspective. 2015;**6**(1):e00392. DOI: 10.1002/brb3.392

[143] Kim TY, Kim SJ, Chung HG, Choi JH, Kim SH, Kang JI. Epigenetic alterations of the BDNF gene in combat-related post-traumatic stress disorder. Acta Psychiatrica Scandinavica. 2017;**135**(2):170-179. DOI: 10.1111/acps.12675

[144] Voisey J, Lawford B, Bruenig D, Harvey W, Morris CP, Young RM, et al. PTSD Initiative. Differential BDNF methylation in combat exposed veterans and the association with exercise. Gene. 2019;**698**:107-112. DOI: 10.1016/j.gene.2019.02.067

[145] Guo JC, Yang YJ, Zheng XA, Jiang XL, Guo M, Wang XD, et al. CpG methylation of brain-derived the neurotrophic factor gene promoter as a potent diagnostic and prognostic biomarker for post-traumatic stress disorder. International Journal of Clinical and Experimental Pathology. 2018;**11**(10):5101-5109

[146] Valente NL, Vallada H, Cordeiro Q, et al. Candidate-gene approach in posttraumatic stress disorder after urban violence: Association analysis of the genes encoding serotonin transporter, dopamine transporter, and BDNF. Journal of Molecular Neuroscience. 2011;**44**:59-67. DOI: 10.1007/s12031-011-9513-7

[147] Lee H, Kang R, Lim S, Paik J, Choi M, Lee M. No association between the brain-derived neurotrophic factor gene Val66Met polymorphism and post-traumatic stress disorder. Stress and Health. 2006;**22**:115-119

[148] Zhang H, Ozbay F, Lappalainen J, Kranzler HR, van Dyck CH, Charney DS, et al. Brain derived neurotrophic factor (BDNF) gene variants and Alzheimer's disease, affective disorders, posttraumatic stress disorder, schizophrenia, and substance dependence. American Journal of Medical Genetics. 2006;**141B**:387-393. DOI: 10.1002/ajmg.b.30332

[149] Hauck S, Kapczinski F, Roesler R, de Moura Silveira E Jr, Magalhaes PV, Kruel LR, et al. Serum brain-derived neurotrophic factor in patients with trauma psychopathology. Progress in Neuro-Psychopharmacology & Biological Psychiatry. 2010;**34**:459-462. DOI: 10.1016/j.pnpbp.2010.01.010

[150] Mojtabavi H, Saghazadeh A, van den Heuvel L, Bucker J, Rezaei N. Peripheral blood levels of brain-derived neurotrophic factor in patients with post-traumatic stress disorder (PTSD): A systematic review and meta-analysis. PLoS One. 2020;**15**(11):e0241928. DOI: 10.1371/journal.pone.0241928

[151] Wu GWY, Wolkowitz OM, Reus VI, Kang JI, Elnar M, Sarwal R, et al. Serum brain-derived neurotrophic factor remains elevated after long term follow-up of combat veterans with chronic post-traumatic stress disorder. Psychoneuroendocrinology. 2021;**134**:105360. DOI: 10.1016/j.psyneuen.2021.105360

[152] Bartel DP. MicroRNAs: Genomics, biogenesis, mechanism, and function. Cell. 2004;**116**:281-297. DOI: 10.1016/s0092-8674(04)00045-5

[153] Varendi K, Kumar A, Härma MA, Andressoo JO. miR-1, miR-10b, miR-155, and miR-191 are novel regulators of BDNF. Cellular and Molecular Life Sciences. 2014;**71**(22):4443-4456. DOI: 10.1007/s00018-014-1628-x. doi: 10.1007/s00018-014-1628-x

[154] Lau AG, Irier HA, Gu J, Tian D, Ku L, Liu G, et al. Distinct 3'UTRs differentially regulate activity-dependent translation of brain-derived neurotrophic factor (BDNF). Proceedings of the National Academy of Sciences of the United States of America. 2010;**107**:15945-15950. DOI: 10.1073/pnas.1002929107

[155] Li YJ, Xu M, Gao ZH, Wang YQ, Yue Z, Zhang YX, et al. Alterations of serum levels of BDNF-related miRNAs in patients with depression. PLoS One. 2013;**8**:e63648. DOI: 10.1371/journal.pone.0063648

[156] Abdolahi S, Zare-Chahoki A, Noorbakhsh F, Gorji A. Review of molecular interplay between neurotrophins and miRNAs in neuropsychological disorders. Molecular Neurobiology. 2022;**59**(10):6260-6280. DOI: 10.1007/s12035-022-02966-5

[157] Maurel OM, Torrisi SA, Barbagallo C, Purrello M, Salomone S, Drago F, et al. Dysregulation of miR-15a-5p, miR-497a-5p and miR-511-5p is associated with modulation of BDNF and FKBP5 in brain areas of PTSD-related susceptible and resilient mice. International Journal of Molecular Sciences. 2021;**22**(10):5157. DOI: 10.3390/ijms22105157

[158] Wan Y, Liu Y, Wang X, Wu J, Liu K, Zhou J, et al. Identification of differential MicroRNAs in cerebrospinal fluid and serum of patients with major depressive disorder. PLoS One. 2015;**10**(3):e0121975. DOI: 10.1371/journal.pone.0121975

[159] Murphy CP, Singewald N. Role of MicroRNAs in anxiety and anxiety-related disorders. Current Topics in Behavioral Neurosciences. 2019;**42**:185-219. DOI: 10.1007/7854_2019_109

[160] Gunasekaran S, Jacob RS, Omkumar RV. Differential expression of miR-148b, miR-129-2 and miR-296 in animal models of schizophrenia-relevance to NMDA receptor hypofunction. Neuropharmacology. 2022;**210**:109024. DOI: 10.1016/j.neuropharm.2022.109024

[161] Ji LL, Ye Y, Nie PY, Peng JB, Fu CH, Wang ZY, et al. Dysregulation of miR-142 results in anxiety-like behaviors following single prolonged stress. Behavioural Brain Research. 2019;**365**:157-163. DOI: 10.1016/j.bbr.2019.03.018

[162] Wang Z, Zhang C, Huang J, Yuan C, Hong W, Chen J, et al. MiRNA-206 and BDNF genes interacted in bipolar I disorder. Journal of Affective Disorders. 2014;**162**:116-119. DOI: 10.1016/j.jad.2014.03.047

[163] Lee SY, Wang TY, Lu RB, Wang LJ, Chang CH, Chiang YC, et al. Peripheral BDNF correlated with miRNA in BD-II patients. Journal of Psychiatric Research. 2021;**136**:184-189. DOI: 10.1016/j.jpsychires.2021.02.018

Chapter 2

The Role of Brain-Derived Neurotrophic Factor in Autism Spectrum Disorder: Current Findings and Future Directions

Mumin Alper Erdogan and Oytun Erbaş

Abstract

Brain-derived neurotrophic factor (BDNF) is a crucial neurotrophic factor that plays an essential role in neuroplasticity and neurodevelopment. Autism spectrum disorder (ASD) is a neurodevelopmental disorder that affects social interaction, communication, and behavior. The relationship between BDNF and ASD has been studied extensively, with conflicting results. While some studies suggest that decreased BDNF levels may contribute to the development of ASD, others do not confirm this finding. The effects of BDNF on synaptic plasticity and cognitive functions have also been investigated, with some studies indicating that BDNF may be associated with impairments in learning, memory, and attention in individuals with ASD. Additionally, physical exercise and cognitive and behavioral therapies may help alleviate ASD symptoms by increasing BDNF levels and enhancing neuroplasticity. Further research is needed to better understand the mechanisms underlying the relationship between BDNF and ASD and to develop more effective treatment strategies for individuals with ASD.

Keywords: BDNF, autism spectrum disorder, neuroplasticity, cognitive functions, therapeutic interventions

1. Introduction

The growth and plasticity of the brain are significantly influenced by the protein BDNF. A neurodevelopmental disorder called Autism spectrum disorder (ASD) causes social and behavioral difficulties. Numerous experts have conducted considerable study on the link between BDNF and ASD.

A neurotrophic factor known as BDNF helps neurons across the central nervous system to survive, develop, differentiate, and function. The nervous system's capacity to adapt to structural and functional changes is known as neuroplasticity. Understanding how BDNF affects neuroplasticity is important for learning, memory, and cognitive functions as well as for understanding the origins and therapies of neurological and neuropsychiatric illnesses.

 IntechOpen

2. The effect of BDNF on neuroplasticity

BDNF plays a critical role in regulating synaptic plasticity processes. Synaptic plasticity can be defined as the strengthening or weakening of synapses, leading to changes in the connections between neurons. BDNF particularly influences the following neuroplasticity processes: [1–3].

1. Synaptogenesis: BDNF promotes the formation of synapses and the development of connections between nerve cells.

2. Dendritic growth and arborization: BDNF supports the development and complexity of neurons' dendritic trees, thereby increasing the number of connections between nerve cells.

3. Synaptic transmission and modulation: BDNF regulates the effectiveness of synaptic transmission and modulation of synapses, which are critical for learning and memory processes.

4. Structural reorganization and neuronal migration: BDNF plays a significant role in the structural organization and reorganization of the brain by regulating the migration and settlement of neurons.

5. Neuronal survival and neuroprotection: BDNF provides protection against neuronal damage by regulating the expression of factors necessary for neurons' survival, growth, and differentiation.

2.1 BDNF and Learning, Memory, and Cognitive Functions

Neuroplasticity is considered a fundamental mechanism in learning and memory processes, and BDNF's effects on these processes are of great importance. BDNF is particularly associated with the following cognitive functions: [2–4].

1. Learning and memory: The hippocampus and prefrontal cortex are particularly where BDNF influences learning and memory processes. As BDNF levels rise, synaptic connections between nerve cells become stronger and more synapses are formed, which enhances the potential for learning and memory. Impaired memory and learning might result from BDNF insufficiency.

2. Processing speed and focus: BDNF is also essential in controlling cognitive processes including processing speed and attentiveness. Levels of BDNF have been demonstrated to be positively correlated with attention and processing speed. Additionally, attention-deficit/hyperactivity disorder (ADHD) and other cognitive function abnormalities have been linked to BDNF insufficiency.

3. Executive functions: BDNF is essential for the development and regulation of executive functions as well as for controlling the prefrontal cortex's functioning. High-level cognitive activities including problem-solving, planning, adaptability, and thinking organizing are included in executive functions. Executive function issues and poor cognitive flexibility may result from BDNF insufficiency.

2.2 BDNF and neurological and neuropsychiatric disorders

The regulation of neuroplasticity and BDNF plays a significant role in the pathogenesis and treatment of neurological and neuropsychiatric disorders. Decreased BDNF levels and impaired neuroplasticity processes have been associated with the following diseases: [2–4].

1. Depression and anxiety: Reduced BDNF levels and impaired neuroplasticity processes are associated with the pathogenesis of depression and anxiety disorders. Antidepressants and other pharmacological treatments are thought to be effective by increasing BDNF levels and supporting neuroplasticity.

2. Schizophrenia: Schizophrenia has been associated with impairments in synaptic function and neuroplasticity processes, and decreased BDNF levels are also observed in this disorder. Antipsychotic drugs used in the treatment of schizophrenia are believed to be effective by increasing BDNF levels and supporting neuroplasticity.

3. Alzheimer's disease and other neurodegenerative conditions: The pathophysiology of Alzheimer's disease, Parkinson's disease, and other neurodegenerative conditions is linked to neuronal damage and compromised neuroplasticity processes. BDNF levels have been found to be decreased in many disorders, and its neuroprotective properties are thought to be a possible target for their therapy.

4. ASD (Autism spectrum disorder): ASD is a neurodevelopmental condition marked by challenges with social interaction and communication as well as confined, monotonous, and stereotyped behaviors. Because BDNF levels are low in ASD, its effects on neuroplasticity are thought to have a possible involvement in the etiology and management of ASD.

2.3 BDNF and neuroplasticity: applications and future research directions

Current research on BDNF and neuroplasticity has provided important insights into the understanding and treatment of neurological and neuropsychiatric disorders [1]. Some important areas of future research that could be focused on in this field include:

1. Drug development: New treatment strategies that regulate BDNF levels and support neuroplasticity processes may be potentially effective in the treatment of neurological and neuropsychiatric disorders. Research in this area should aim to discover new pharmacological compounds and enhance the effectiveness of current treatments [2].

2. Behavioral and lifestyle interventions: Research could investigate the effects of physical activity, diet, and other lifestyle factors on increasing BDNF levels and neuroplasticity naturally. Such interventions could provide a complementary approach for the prevention and treatment of neurological and neuropsychiatric disorders [3].

3. Neurodevelopmental processes and aging: Research on BDNF and neuroplasticity could contribute to a better understanding of neurodevelopmental processes

in early life and cognitive and neurological changes associated with aging. Such studies could lead to the development of specific strategies to support neuroplasticity at different stages of life [4].

4. Personalized medicine: Research on BDNF and neuroplasticity could help determine the impact of individual genetic and environmental factors on disease risk and treatment effectiveness. Such information could contribute to the development of personalized treatment approaches and more effective management of neurological and neuropsychiatric disorders [5, 6].

Future research on BDNF and neuroplasticity has a great potential to provide a better understanding of how neuronal and synaptic functions change in different disease states and stages of life. Specifically, gaining more knowledge on how BDNF and neuroplasticity mechanisms interact and influence each other could lead to the development of more effective treatment strategies and better management of neurological and neuropsychiatric disorders. Progress in this field could play an important role in improving patients' quality of life and contributing to public health.

3. BDNF's biological and functional properties and effects

3.1 Biological properties of BDNF

BDNF is a protein produced in nerve cells in the brain and plays an important role in many biological processes such as neurodevelopment and synaptic plasticity [7]. BDNF is converted from proBDNF, a protein synthesized in brain cells and sent to neurons, to mature BDNF (mBDNF) by proteolytic cleavage [8]. BDNF is particularly expressed in brain regions such as the hippocampus, prefrontal cortex, striatum, and amygdala [9, 10].

3.2 Functional properties and effects of BDNF

BDNF is a protein that affects communication between neurons in the brain and plays an important role in many biological processes such as neurodevelopment and synaptic plasticity [6]. BDNF promotes the growth and healthy development of neurons. Moreover, BDNF strengthens synaptic connections between neurons and supports the formation of new synaptic connections [7, 8]. BDNF is also important for learning and memory and plays a role in memory formation [9]. BDNF also plays an important role in regulating stress response and mood [10].

The effects of BDNF are mediated through receptors. BDNF binds to a receptor called TrkB to promote the growth and healthy development of neurons [6]. Additionally, TrkB receptor strengthens synaptic connections between neurons and supports the formation of new synaptic connections [11]. TrkB receptors are also responsible for the effects of BDNF on learning and memory formation [12].

As BDNF plays a significant role in regulating nervous system functions, BDNF levels can vary in many diseases associated with processes such as neurodevelopment and synaptic plasticity [13]. Therefore, BDNF levels are also used as a potential biomarker for the pathophysiology, diagnosis, and treatment of neuropsychiatric disorders [14].

3.3 BDNF and neurodevelopment

BDNF promotes the growth, migration, and differentiation of nerve cells during neurodevelopment. BDNF also assists nerve cells in forming the proper connections. A deficiency in BDNF can result in errors in neurodevelopment and the failure of neurons to make the proper connections [11]. The effects of BDNF on neurodevelopment have been studied extensively in relation to neurodevelopmental disorders [12].

3.4 BDNF gene expression

BDNF gene expression is necessary for the production of BDNF protein. The BDNF gene can be expressed by neurons and other cell types [13]. BDNF gene expression is influenced by many factors, such as activity, stress, and neurodevelopmental processes [14]. BDNF gene expression has been studied extensively in relation to neurodevelopmental disorders and other brain diseases [9].

3.5 BDNF's roles in different brain regions

The roles of BDNF vary in different regions of the brain. In the hippocampus, BDNF is involved in learning and the formation of memories [15]. BDNF also plays a crucial role in regulating stress response and emotion in the prefrontal cortex [16]. Additionally, other brain regions such as the striatum and amygdala also rely on BDNF for proper functioning [17].

4. Autism spectrum disorder

Autism spectrum disorder (ASD) is a condition that stems from the interplay of both genetic and environmental elements and affects neurodevelopment. ASD is defined by symptoms such as challenges with social interactions, communication deficits, and repetitive and restricted behavior patterns [18].

1. Pathophysiology of ASD:

   The exact cause of autism spectrum disorder (ASD) remains unclear, but it is thought to be the result of a complex interplay between genetic, epigenetic, and environmental factors affecting the development and function of the brain [19]. Many researchers suggest that ASD arises from dysfunctions in brain development and function [20]. Brain development is related to the proper migration, differentiation, and connection of neurons. In addition, the proper formation and function of synaptic connections between nerve cells is also important [21].

2. Relationship between ASD Neurodevelopment and BDNF

   The relationship between the neurodevelopmental abnormalities in ASD and BDNF has been studied by many researchers. It has been found that BDNF levels are decreased in individuals with ASD, especially in those with low functional levels on the autism spectrum [22, 23]. In contrast, BDNF receptor levels in individuals with ASD are normal or increased [24].

3. BDNF and ASD Symptoms:

   BDNF is a key factor in synaptic plasticity and neurodevelopment, and a decrease in its levels may be linked to the symptoms of ASD. Specifically, a decrease in BDNF levels can result in an increase in social interaction difficulties and repetitive behaviors among individuals with ASD [25]. Additionally, the decreased BDNF levels observed in individuals with ASD have been linked to emotional disorders and increased obsessive-compulsive behaviors [26].

4. BDNF, ASD Treatment, and Medications:

   BDNF may be a potential target in the treatment of ASD. The neurodevelopmental effects of BDNF can be used to improve brain function in individuals with ASD [27]. Increasing BDNF levels may increase synaptic plasticity and reduce ASD symptoms. Therefore, drugs that increase BDNF levels are being investigated as a potential strategy in the treatment of ASD [28].

ASD and BDNF Gene Expression: Decreased expression levels of the BDNF gene in individuals with ASD may be associated with developmental dysfunctions. Some studies have shown that BDNF gene expression levels may be decreased in individuals with ASD [29, 30].

BDNF and ASD Medications: Drugs that increase BDNF levels are being evaluated as a potential strategy for ASD treatment [31]. For example, antidepressant drugs such as selective serotonin reuptake inhibitors (SSRIs) are thought to reduce ASD symptoms by increasing BDNF levels [32]. Additionally, BDNF agonists are being investigated as a potential treatment strategy for reducing ASD B symptoms [33].

4.1 Clinical features, diagnosis, and treatment of autism spectrum disorder

Early childhood is when autism spectrum disorder (ASD) first appears. ASD is characterized by challenges with social interaction and communication as well as limited and repetitive behavioral patterns. The three core characteristics of ASD, as defined by the DSM-5, include difficulties with social interaction and communication, as well as restricted interests and repetitive behaviors. These signs can be mild to severe and last a person their entire life [18]. A thorough assessment of a child's behavioral traits, such as social interaction, language and communication abilities, repetitive habits, and interests, can lead to the diagnosis of ASD. Specialists frequently utilize standardized tests and autism screening instruments to make their diagnoses. However, identifying ASD cannot be done with a single test or clear indication. Input from a child's family, teachers, and other healthcare experts may also be included in a thorough review [34]. Multidisciplinary therapy is necessary for ASD. A child's treatment frequently starts as early as feasible and lasts their entire lives. Education, speech and language therapy, behavior therapy, family counseling, and medication are all possible treatment modalities. Children can have better results with early diagnosis and treatment [35].

5. BDNF and autism spectrum disorder

BDNF is a member of the neurotrophic factor family and is critical for neurological functions such as neuroplasticity and neurogenesis. BDNF functions as a protein that regulates the growth, maturation, survival, and synaptic plasticity of neurons [36].

Changes in BDNF levels in individuals with autism spectrum disorder (ASD) may contribute to the pathophysiology of ASD. Many researchers have found evidence that changes in BDNF levels may be associated with ASD. Some studies have shown that BDNF levels are lower in individuals with ASD and that these lower levels are associated with ASD symptoms [37]. However, other studies suggest that normal BDNF levels may be associated with ASD. For example, one study found that individuals with ASD had normal BDNF levels compared to a control group, but differences in the regional distribution of BDNF in the brain may contribute to ASD symptoms [38]. Additionally, genetic variations in the BDNF gene have been investigated in individuals with ASD. One study found that certain variations in the BDNF gene were associated with an increased risk of ASD [39]. However, another study found that these variations in the BDNF gene were not associated with ASD [40]. The relationship between BDNF and ASD is not yet fully understood and further research is needed in this area. Taken together, the evidence discussed suggests that BDNF may have an important role in the pathophysiology of ASD, although the precise nature of this role warrants further research.

The relationship between ASD and BDNF may be important for the pathophysiology of ASD, and further research in this area is needed. Many researchers have shown that BDNF levels are decreased in individuals with ASD and that these low levels are associated with ASD symptoms. However, other studies suggest that normal BDNF levels may also be associated with ASD.

BDNF levels may be used as a potential therapeutic target to alleviate ASD symptoms. A study has shown that BDNF deficiency in mice leads to ASD-like symptoms and that BDNF infusion can reverse these symptoms. This study suggests that BDNF may be a potential agent for ASD treatment.

In conclusion, while the relationship between ASD and BDNF is not yet fully understood, it is known that BDNF is critical for neurological functions such as neuroplasticity and neurogenesis and may play a role in ASD pathophysiology. The diagnosis and treatment of ASDB require a multidisciplinary approach, and early diagnosis and treatment may help achieve better outcomes. BDNF levels may be used as a potential therapeutic target in ASD treatment.

6. BDNF's role in the pathophysiology of autism spectrum disorder and clinical outcomes

Although the exact role of BDNF in the pathophysiology of ASD is still not fully understood, studies in this area have made significant progress. Changes in BDNF levels have been shown to be associated with ASD, and BDNF receptors and signaling pathways are also thought to play an important role in ASD pathophysiology.

6.1 Changes in BDNF levels

Changes in BDNF levels may be related to ASD pathophysiology. Some studies have shown low levels of BDNF in individuals with ASD [41–45]. These low levels have also been suggested to be associated with ASD symptoms [42]. Some research suggests that changes in BDNF levels are associated with factors that affect BDNF production in the brain. For example, one study showed that maternal antibodies inhibited BDNF production in fetal mice, resulting in ASD-like symptoms [46]. Another study showed that early-life stress in mice resulted in decreased BDNF levels, which were associated with ASD-like symptoms [47].

6.2 BDNF receptors and signaling pathways

BDNF's effects are mediated by tropomyosin receptor kinase B (TrkB) receptors, which are high-affinity receptors on the cell surface [48]. Activation of TrkB receptors by BDNF affects a series of signaling pathways critical for neurological functions such as neuroplasticity and neurogenesis [49]. BDNF activates signaling pathways that affect many neurological functions, including neurotransmitter release, synaptic plasticity, cell proliferation, and cell differentiation, through TrkB receptors [50, 51]. Therefore, the role of TrkB receptors and these signaling pathways in the pathophysiology of ASD is also being investigated.

Some studies have shown that TrkB receptor levels are low in individuals with ASD [51]. These low levels may contribute to ASD pathophysiology by reducing the effects of BDNF. In addition, other components of BDNF signaling pathways may also play a role in the pathophysiology of ASD. For example, a study showed that the phosphatidylinositol 3-kinase (PI3K)/AKT signaling pathway is involved in BDNF's neuroprotective effects and may also play an important role in the pathophysiology of ASD [52].

6.3 Clinical implications of BDNF and ASD

While the exact role of BDNF in ASD pathophysiology is not fully understood, research in this area has made significant progress. Changes in BDNF levels have been shown to be associated with ASD, and BDNF receptors and signaling pathways may also play an important role in ASD pathophysiology.

Several studies have shown that low BDNF levels are associated with ASD symptoms [48, 49]. It has also been suggested that an increase in BDNF levels may alleviate ASD symptoms [50]. Additionally, BDNF levels could be a potential therapeutic target for ASD treatment. Some studies have shown that BDNF agonists, in particular, may have a potential role in alleviating ASD symptoms [51, 52].

However, further research is needed to fully understand the potential use of BDNF in ASD treatment. The side effects of BDNF, especially with long-term use, are not yet fully understood and require careful investigation.

6.4 Relationship between BDNF levels and severity of ASD

Studies on individuals with ASD indicate that BDNF levels are associated with the severity of the disorder. Specifically, low levels of BDNF have been linked to more severe ASD symptoms [53]. Various studies have reported that plasma and serum BDNF levels in individuals with ASD are lower compared to those without ASD [54]. However, it is believed that changes in BDNF levels may vary across different subtypes of ASD [55]. Furthermore, a positive correlation has been reported between BDNF levels and social functioning [56]. This relationship suggests that an increase in BDNF levels is paralleled by improvement in social skills. These findings suggest that BDNF plays an important role in regulating the neurobiological mechanisms and modulating symptom severity in ASD.

6.5 The effect of BDNF on cognitive and social functions in individuals with ASD

BDNF plays an important role in the development and regulation of cognitive and social functions. Studies conducted in individuals with ASD have shown that BDNF levels affect learning, memory, language skills, and social skills [57]. For example, a study conducted in children with ASD found a positive correlation between BDNF levels and language development and social skills [58]. These results indicate that BDNF is an important modulator of language and social skill development in individuals with ASD. Increasing BDNF has been associated with improvements in cognitive function and social skills [59]. Therefore, increasing BDNF levels is considered a potential treatment approach for improving cognitive and social functions in individuals with ASD [54]. Pharmacological treatments and lifestyle changes targeting BDNF, particularly BDNF agonists, are evaluated as promising methods to enhance cognitive and social functions in individuals with ASD. These treatments may target neurotraumatic factors, synaptic plasticity, and neurogenesis processes to increase BDNF levels. However, further research is needed to fully understand the effects of BDNF on cognitive and social functions in individuals with ASD. Future studies should focus on evaluating the efficacy and safety of treatment strategies targeting BDNF and improving our understanding of the complex interactions between BDNF and ASD [2, 60].

7. BDNF genetic and epigenetic regulations: Their association with autism spectrum disorder (ASD)

7.1 BDNF genetic regulations and ASD

1. BDNF Polymorphisms: Various polymorphisms in the BDNF gene have been associated with the risk of ASD. In particular, the Val66Met (rs6265) polymorphism is the most commonly reported BDNF polymorphism in ASD. This polymorphism can affect the conversion and release of BDNF from pro-BDNF to mature BDNF, leading to disrupted synaptic plasticity and neuronal communication [61].

2. ASD Severity and BDNF Genetic Variations: BDNF polymorphisms have also been linked to the severity and clinical features of ASD. For example, in addition to the Val66Met polymorphism, other polymorphisms in the BDNF gene (e.g. rs2049046 and rs11030104) have been associated with ASD severity and clinical characteristics [62].

7.2 BDNF epigenetic regulations and ASD

1. DNA Methylation: DNA methylation levels in the promoter region of the BDNF gene play a crucial role in ASD. Abnormal DNA methylation levels in the BDNF promoter region have been reported in individuals with ASD. These abnormal methylation levels can cause changes in BDNF gene expression and disrupt synaptic plasticity and neuroplasticity, which are key neurobiological mechanisms underlying ASD [63].

2. Histone Modifications: Histone modifications of the BDNF gene can also impact ASD. In particular, histone acetylation and methylation levels can regulate BDNF gene expression and influence synaptic plasticity and neuroplasticity [64]. Changes in histone modifications of the BDNF gene have been observed in individuals with ASD, which can cause disruptions in synaptic function at the neurobiological level.

Genetic and epigenetic regulations in the BDNF gene play an important role in the neurobiological basis of autism spectrum disorder (ASD). BDNF gene polymorphisms and epigenetic regulations can affect synaptic plasticity and neuroplasticity, and therefore have been associated with ASD risk and severity. Understanding the role of BDNF's genetic and epigenetic regulations in the etiology of ASD may contribute to the development of new intervention and treatment strategies.

8. The role of BDNF in the neurodevelopment, neuroplasticity, and cognitive functions of autism spectrum disorder

BDNF is a crucial neurotrophic factor for the survival, development, and function of neurons [65]. Additionally, it has a significant impact on neuroplasticity and neurodevelopment, and has been linked to neurological disorders such as autism spectrum disorder. While some studies suggest a decrease in BDNF levels in individuals with ASD, others have not confirmed this finding [32, 66, 67]. Synaptic plasticity is an important mechanism for neurons to modify their ability to communicate with each other, and is essential for neurodevelopment and learning processes. BDNF's effects on synaptic plasticity have been associated with neurological disorders like autism spectrum disorder, with some studies indicating that synaptic plasticity may be impaired in individuals with ASD [68, 69]. The effects of BDNF on cognitive functions have also been investigated. A decrease in BDNF levels in individuals with autism spectrum disorder may lead to cognitive impairments, with some studies suggesting that memory, learning, and attention may be affected in individuals with ASD [61, 70].

Based on the literature findings regarding the role of BDNF in the neurodevelopment, neuroplasticity, and cognitive functions of individuals with autism spectrum disorder (ASD), changes in BDNF levels may play a role in the pathophysiology of ASD, but the exact mechanism is still not fully understood. BDNF deficiency, as suggested by some studies, can affect ASD in several ways. For example, BDNF deficiency can affect the maturation and function of synapses in neurons during neurodevelopment. However, BDNF deficiency is thought to be particularly effective on synaptic plasticity and cognitive functions in brain regions such as the hippocampus and amygdala. BDNF deficiency may also be associated with fundamental symptoms of ASD, such as social behavior and communication. Some studies suggest that BDNF deficiency could help develop various treatments to alleviate ASD symptoms. For instance, treatments that increase BDNF levels have been shown to support the development of social interaction, language skills, and cognitive functions in children with ASD.

9. BDNF and neuroinflammation in ASD

Neuroinflammation is a factor associated with the pathogenesis of ASD. BDNF's anti-inflammatory properties and neuroprotective effects may play a role in managing neuroinflammation in ASD.

9.1 Neuroinflammation and ASD

Neuroinflammation is a process involving inflammatory responses and release of inflammatory mediators by nervous system cells. In the context of ASD pathogenesis, possible mechanisms of neuroinflammation include immune cell activation, cytokine and chemokine production, oxidative stress, and neurotransmitter imbalances.

Neuroinflammation in ASD is associated with activation of immune cells such as microglia and astrocytes in the brain. These activated cells produce proinflammatory cytokines and chemokines, which contribute to the maintenance of neuroinflammation and disruption of synaptic function.

9.2 The anti-inflammatory and neuroprotective effects of BDNF

BDNF is one of the neurotrophic factors that are important for the survival, growth, and differentiation of nerve cells. The anti-inflammatory and neuroprotective properties of BDNF may contribute to the management of neuroinflammation in ASD by reducing inflammation and protecting nerve cells. BDNF can regulate inflammatory processes and decrease the activation of immune cells. As an instance, BDNF could decrease neuroinflammation by promoting the generation of anti-inflammatory cytokines, including interleukin-10 (IL-10) and transforming growth factor-beta (TGF-β). Furthermore, BDNF may alleviate the effects of neuroinflammation by reducing oxidative stress and regulating neurotransmitter balance. BDNF may contribute to the preservation of synaptic function and maintenance of neuroplasticity, thereby affecting the development and severity of ASD [71].

9.3 Modulation of BDNF and neuroinflammation in ASD

Studies investigating the potential role of BDNF in managing neuroinflammation in ASD indicate that this neurotrophic factor may contribute to reducing inflammation and protecting nerve cells [72–74]. For example, the effect of BDNF on astrocytes, which play an important role in regulating neuroinflammation, may affect inflammatory processes in ASD [75]. In addition, interventions targeting BDNF may have positive effects on reducing neuroinflammation and protecting nerve cells in individuals with ASD. Pharmacological agents or gene therapy methods used to increase BDNF levels may contribute to managing neuroinflammation in ASD and alleviating its symptoms [76]. In conclusion, the role of neuroinflammation in the relationship between BDNF and ASD is an important area of research for better understanding the potential impact of this neurotrophic factor on the pathogenesis and treatment of ASD. Future studies examining the modulation of neuroinflammation and the preservation of synaptic function in ASD by BDNF may contribute to the development of new and effective treatment strategies. These investigations are of great importance for the development of methods that may be used for the treatment of ASD and other neurodevelopmental disorders by improving the understanding of the anti-inflammatory and neuroprotective properties of BDNF.

10. BDNF's potential effects on treatment of autism spectrum disorder

10.1 BDNF and ASD treatment

The effect of BDNF on neuroplasticity and synaptic function may play an important role in alleviating ASD symptoms. Pharmacological and behavioral approaches that increase BDNF levels and enhance neuroplasticity can be used in ASD treatment.

10.2 Pharmacological approaches

1. Antidepressants: Antidepressants such as selective serotonin reuptake inhibitors (SSRIs) can enhance neuroplasticity by increasing BDNF levels [75]. Therefore, SSRIs and other antidepressants have the potential to modulate BDNF levels in ASD treatment.

2. Neurotrophic Factor Modulators: Neurotrophic factor modulators that affect BDNF can alleviate ASD symptoms by increasing BDNF levels and supporting neuroplasticity [76]. Such drugs can be effective in ASD treatment by promoting the survival and growth of nerve cells.

10.3 Behavioral approaches

1. Physical activity: By raising BDNF levels and promoting neuroplasticity, physical activity can reduce the symptoms of ASD [77]. Therefore, engaging in regular physical exercise can significantly enhance an ASD person's quality of life and ability to adjust to social situations.

2. Cognitive and behavioral therapies: These treatments have the potential to help people with ASD become more socially and communicationally adept. These treatments can help to reduce ASD symptoms by regulating BDNF levels and neuroplasticity [78].

In conclusion, BDNF and ASD treatment is a promising research area for alleviating ASD symptoms using a combination of pharmacological and behavioral approaches. By increasing BDNF levels and promoting neuroplasticity, these approaches can enhance the quality of life and social adaptation of individuals with ASD. Furthermore, treatment strategies that increase BDNF levels can provide further insights into the pathophysiology and treatment of ASD by elucidating their effects on neuroplasticity and synaptic function. BDNF plays a significant role in regulating neurodevelopment, synaptic plasticity, and cognitive function. Therefore, BDNF-targeted therapies may have potential benefits for the treatment of autism spectrum disorder.

10.4 BDNF targeted treatment options

BDNF targeted treatment options include both pharmacological and non-pharmacological approaches. Pharmacological treatments include medications such as antidepressants, antipsychotics, and sodium valproate. Some studies have shown that sodium valproate can reduce symptoms of autism spectrum disorder by increasing BDNF levels [79]. Antidepressants may be effective in treating comorbid symptoms commonly seen in autism spectrum disorder, such as obsessive-compulsive disorder and depression. Antipsychotics are used to treat disruptive behaviors in autism spectrum disorder. Non-pharmacological treatments include exercise, diet, cognitive therapy, and cognitive-behavioral therapy. Exercise, in particular, is thought to increase neurodevelopment and synaptic plasticity by leading to an increase in BDNF levels [80]. Diet can also be helpful in treating symptoms of autism spectrum disorder. For example, one study showed that omega-3 fatty acids can reduce hyperactivity symptoms in autism spectrum disorder [81]. Cognitive therapy and cognitive-behavioral therapy are effective treatment options for symptoms such as anxiety and depression in autism spectrum disorder.

10.5 Possible side effects of using BDNF

Potential side effects of BDNF-targeted treatments include headaches, sleep disturbances, and sexual dysfunction with antidepressants; movement disorders and weight gain with antipsychotics; and impaired liver function with sodium valproate [82, 83].

These studies suggest that increasing BDNF levels may help improve symptoms of autism spectrum disorder. However, the effectiveness of BDNF-targeted treatments is still being investigated, and further research is needed. In this section, BDNF-targeted treatment options, both pharmacological and non-pharmacological, as well as possible side effects of BDNF use will be discussed.

BDNF-targeted treatments include BDNF agonists and BDNF enhancers. BDNF agonists increase the effects of BDNF by binding to BDNF receptors, while BDNF enhancers increase BDNF production and enhance the response of neurons to BDNF. Animal studies have shown that BDNF agonists may be effective in improving symptoms of autism spectrum disorder. However, the effectiveness of these treatments in humans is still being investigated.

Pharmacological treatments that can increase BDNF levels include antipsychotics, antidepressants, and psychostimulants. However, the side effects of these medications should also be considered. In particular, metabolic side effects of antipsychotics are a significant concern for their use in children and adolescents.

Non-pharmacological treatments that can increase BDNF levels include physical activity, exercise, meditation, and therapy. For example, physical activity and exercise have been shown to increase BDNF levels and enhance neuroplasticity. Similarly, stress management techniques such as meditation and therapy have been shown to increase BDNF levels.

Possible side effects of BDNF-targeted treatments may include neurotoxicity due to excessive BDNF increases and BDNF's pro-inflammatory effects. Therefore, these treatments should be carefully managed.

In conclusion, BDNF-targeted treatments may have potential benefits for autism spectrum disorder. However, the side effects and effectiveness of treatment options need to be considered. Further research is needed to ensure the appropriate use of BDNF-targeted treatments.

11. Recent research findings and future research directions on BDNF

An essential neurotrophin known as BDNF is involved in the cognitive, neurodevelopmental, and neuroplastic aspects of autism spectrum disorder. More details on the function of BDNF in the pathophysiology of autism spectrum disorder have come to light recently. Future study is required since it is yet unknown how BDNF affects the therapy of autism spectrum disorder.

11.1 Control of BDNF gene expression

The usage of BDNF in the treatment of autism spectrum disorder can be improved by managing the expression of the BDNF gene. More investigation is required, in particular, on how the BDNF gene-associated SNPs affect the likelihood of developing autism spectrum disorder. A correlation between BDNF polymorphisms and autism spectrum disorder was discovered in one study [84], however further investigation is required to fully understand this correlation.

The usage of BDNF in the treatment of autism spectrum disorder can be improved by managing the expression of the BDNF gene. It is necessary to do additional study on the nature of the association between BDNF polymorphisms and autism spectrum disorder in order to better understand the processes that enhance or decrease BDNF gene expression. According to one study in this field, people with autism spectrum disorder have changed gene regulatory regions that boost the expression of the BDNF gene [85]. This finding raises the possibility that the pathophysiology of autism spectrum disorder may include the control mechanisms of BDNF gene expression.

11.2 Examination of BDNF receptors and signaling pathways

The effects of BDNF are dependent on the activation of BDNF receptors on the cell surface. Therefore, examining the BDNF receptors and signaling pathways may help to better understand the effects of BDNF in the treatment of autism spectrum disorder. One study showed that the effects of BDNF are mediated through the activation of TrkB receptors [86]. However, the subtypes of these receptors and the exact workings of the signaling pathways are still unclear. BDNF affects synaptic plasticity and neurodevelopment through the TrkB receptor. Therefore, a better understanding of the effects of the TrkB receptor and BDNF signaling pathway on the pathophysiology of autism spectrum disorder is needed. One study showed that BDNF increased social behavior through activation of the TrkB receptor and restored normal social behavior in mice with social behavior deficits, which are also present in autism spectrum disorder patients [87]. These results suggest that the TrkB receptor and BDNF signaling pathway may have a significant impact on symptoms of autism spectrum disorder, such as social behavior.

11.3 Understanding the effects of BDNF on behavioral and social functions

A better understanding of the effects of BDNF on behavioral and social functions may assist in the development of BDNF-targeted therapies for autism spectrum disorder (ASD) treatment. Specifically, the effects of BDNF on social functions are still not clear and further research is needed in this area. One study has shown that BDNF treatment improved social learning and increased social memory [88].

These results suggest that BDNF may play a significant role in regulating social functions and the effects of BDNF on behavioral and social functions are seen as a potentially useful area for ASD treatment. BDNF is considered a potential target for treating symptoms of ASD, such as social function impairment, especially social function disorder.

Many studies have demonstrated the positive effects of BDNF on social learning and social memory. For example, one study showed that BDNF application improved social learning and increased social memory [65]. The effects of BDNF are thought to be useful for treating symptoms of ASD, such as social function disorder observed in ASD.

A better understanding of the effects of BDNF on social functioning is important for the development of BDNF-targeted treatments. To do this, more research is needed to understand the role of BDNF in regulating social functioning, particularly its effects on processes such as social learning, processing, and memory. Such studies can help us better understand how effective BDNF-targeted treatments may be in treating symptoms such as social dysfunction in autism spectrum disorder.

11.4 Future perspectives in BDNF and ASD research

In autism spectrum disorder (ASD) and brain-derived neurotrophic factor (BDNF) research, future studies are expected to focus on developing more comprehensive and effective strategies for understanding and treating the disease. Here are some important areas related to these perspectives:

1. Modulation methods for BDNF levels: Future research should focus on discovering ways to modulate BDNF levels. This is considered a potential treatment approach for improving cognitive and social functions in individuals with ASD [89]. It is important to increase the number of clinical studies evaluating the efficacy and safety of BDNF-targeted pharmacological treatments, neuromodulation techniques, and lifestyle changes [90].

2. BDNF and subtypes of ASD: Better understanding of changes in BDNF levels among different subtypes of ASD is needed [32]. Studies examining BDNF levels and mechanisms specific to ASD subtypes can contribute to the development of diagnosis and treatment strategies [91].

3. Epigenetic regulation of BDNF: Studies focusing on the role of epigenetic mechanisms that affect BDNF gene expression and activity in ASD should be increased [70]. Epigenetic regulators such as DNA methylation, histone modifications, and microRNAs can play an important role in the pathophysiology of ASD and offer potential therapeutic targets [92].

4. BDNF and neuroinflammation in ASD: The number of studies investigating the interaction between BDNF and neuroinflammation in ASD should be increased [93]. Understanding the role of BDNF in regulating neuroinflammation and modulating mechanisms related to the immune system in ASD can help us better understand the neurobiological basis of ASD and potential treatment approaches [94].

5. BDNF and synaptic plasticity in ASD: Synaptic plasticity plays an important role in learning and memory processes. Studies investigating the effects of BDNF on synaptic plasticity in ASD should be increased [95]. These studies can shed light on treatment strategies for improving cognitive and social skills in individuals with ASD [96].

6. BDNF and neurogenesis and gliogenesis in ASD: BDNF is important for the development and function of neurons and glial cells. Studies examining the effects of BDNF on neurogenesis and gliogenesis in ASD should be increased [97]. These studies can provide more information about the role of neuronal and glial cells in the pathophysiology of ASD and help develop new treatment approaches [98].

7. Interactive factors with BDNF and ASD: Given the complex nature of ASD, it is important to identify other factors that interact with BDNF. Studies investigating how genetic, environmental, and lifestyle factors affect BDNF levels and the pathophysiology of ASD can fill gaps in knowledge in this field [99].

8. Early diagnosis and prognosis of ASD with BDNF: The number of studies investigating the use of BDNF levels as a potential biomarker for early diagnosis and prognosis of ASD should be increased [100]. Early diagnosis and prognosis are important for initiating effective interventions in a timely manner and improving outcomes [101].

In summary, future perspectives in BDNF and ASD research should focus on comprehensive and innovative studies that will fill the gaps in knowledge and contribute to the development of more effective diagnosis and treatment methods for individuals with ASD. These studies will help us better understand the neurobiological basis of ASD and develop effective treatment strategies.

11.5 Personalized ASD treatment and BDNF

Personalized treatment approaches aim to improve the quality of life and functionality of individuals with autism spectrum disorder (ASD) by offering customized treatment plans based on each individual's genetic, biochemical, and environmental factors. Brain-derived neurotrophic factor (BDNF) can be considered an important target in personalized ASD treatment.

Firstly, identifying BDNF levels and genetic variations can help in selecting appropriate treatment methods based on individual differences. Studies examining BDNF levels and interactive factors can contribute to optimizing treatment options specific to the needs and sensitivities of individuals with ASD.

In addition, pharmacological and lifestyle interventions targeting BDNF can be used in personalized ASD treatment. For example, drugs that increase BDNF levels and support synaptic plasticity can be evaluated as a potential treatment to improve the cognitive and social skills of individuals with ASD, taking into account individual differences. Lifestyle interventions, especially regular physical activity and appropriate nutrition, can help increase BDNF levels and improve the quality of life and functionality of individuals with ASD.

In conclusion, knowing the precise functions of BDNF in ASD and using this information to individualized treatment plans will help to create more successful and focused therapies for people with ASD. Future studies should investigate the relationship between BDNF and the underlying causes of ASD, the variables that control BDNF levels, and the efficacy of BDNF-targeting therapies. Examining BDNF levels and effects in various ASD subtypes and individual variations can also help with the creation of more sensitive and efficient treatment approaches because of the varied character of ASD.

12. Prevention of neurodevelopmental disorders and policies related to ASD

Understanding the relationship between BDNF and ASD can contribute to the prevention of neurodevelopmental disorders and the development of policies and strategies for individuals with ASD. In this context, the following steps are recommended:

1. Increasing Awareness: Raising awareness about the relationship between ASD and BDNF can help the community understand and support the lives of individuals with ASD. This can be achieved through educational programs, public awareness campaigns, and media efforts.

2. Early Intervention and Education Programs: Given that genetic and epigenetic regulations in the BDNF gene may affect the risk and severity of ASD in early life, the importance of early intervention and education programs should be emphasized. These programs should aim to improve the social, communication, and cognitive skills of children with ASD and enhance their quality of life.

3. Control of Environmental Factors: Policies and strategies should be developed to reduce the risk of ASD by considering the impact of environmental factors on epigenetic regulations in BDNF. This may include measures such as preventing exposure to environmental toxins during pregnancy and early life and promoting healthy lifestyle choices.

4. Research and Treatment Development: Further research should be conducted on the relationship between BDNF and ASD, and this information should be used to develop new and effective treatment strategies. In addition, continuous efforts should be made to evaluate and improve the effectiveness of existing treatment approaches.

5. Policy and Legal Regulations: Policies and legal regulations should be established and implemented to improve the lives of individuals with ASD. This should include policies that support the rights and opportunities of individuals with ASD in areas such as education access, employment opportunities, and social services.

In conclusion, the better understanding of the relationship between BDNF and ASD provides important insights into the prevention and management of neurodevelopmental disorders. Specifically, studies on the genetic and epigenetic regulations of BDNF offer new perspectives on the etiology and treatment of ASD. In the future, it is important to conduct more detailed research on the relationship between BDNF and ASD and apply this knowledge to develop effective policies and strategies. This approach can contribute to improving the quality of life of individuals with ASD and enhancing the general ability of society to cope with neurodevelopmental disorders.

13. Conclusions

In conclusion, this section discussed the current scientific literature on the relationship between BDNF and ASD. BDNF was highlighted as an important protein in neuronal functions such as synaptic plasticity, neurogenesis, and gliogenesis, and therefore, it has significant importance in the pathophysiology and treatment of ASD. The role of BDNF in the specificity of ASD and the relationship between individualized ASD treatment and BDNF were also addressed.

BDNF has emerged as a possible target for the therapy of autism spectrum disorder (ASD), as it is a protein that is crucial for neurodevelopment, synaptic plasticity, and cognitive skills. According to recent studies, BDNF levels are linked to ASD symptoms. To ascertain the efficacy and safety of BDNF-targeted therapies, more study is necessary.

BDNF levels have been found to be low in patients with ASD, making BDNF-targeted treatments a potential target for the treatment of ASD. Pharmacological treatment options include antidepressants, antipsychotics, and sodium valproate. Some studies have shown that sodium valproate can increase BDNF levels and reduce symptoms of ASD. However, further research is needed to determine the effectiveness and safety of these treatments.

Non-pharmacological treatment options include exercise, nutrition, and therapy options. Particularly, exercise can increase BDNF production and help reduce symptoms in children with autism spectrum disorder. Cognitive therapy and cognitive-behavioral therapy are also effective treatment options for symptoms such as anxiety and depression in autism spectrum disorder. However, further research is needed on the effects of these non-pharmacological treatments on BDNF levels.

Controlling BDNF gene expression may help in developing the use of BDNF in autism spectrum disorder treatment. Examining BDNF receptor and signaling pathways can also play an important role in developing BDNF-targeted treatments. For example, it has been shown that activation of BDNF's TrkB receptors enhances social behavior and restores normal social behavior in mice with social behavior deficits similar to those seen in autism spectrum disorder patients.

Pharmacological and non-pharmacological options for BDNF-targeted treatments include antidepressants, antipsychotics, sodium valproate, exercise, diet, cognitive therapy, and cognitive-behavioral therapy. The side effects of these treatment options should also be taken into consideration.

The potential effects of BDNF-targeted treatments include increased neurodevelopment and synaptic plasticity, reduced symptoms, and improved behavioral and social functioning in individuals with autism spectrum disorder (ASD). However, the relationship between BDNF and ASD is not yet fully understood, and further research is needed. Understanding the relationship between BDNF and ASD could have significant benefits for clinical and research applications. Specifically, using BDNF levels and genetic variations in the diagnosis and prognosis of ASD could provide opportunities for early intervention and support. Additionally, BDNF-targeted treatment approaches could contribute to the development of potential therapies aimed at improving cognitive and social skills in individuals with ASD. Lastly, evaluating BDNF levels and interactive factors in individualized ASD treatment could provide optimized treatment options tailored to each individual's unique needs and sensitivities. Future research focusing on BDNF gene expression control, BDNF receptors and signaling pathways, and better understanding the effects of BDNF on behavioral and social functioning could help develop BDNF-targeted treatments for use in ASD.

There are some limitations to consider in BDNF and ASD research, as well as suggestions for future studies. Firstly, many current studies may not fully reflect the heterogeneous nature of ASD and may overlook the relationships between different ASD subtypes and individual differences in BDNF levels and effects. Therefore, future research should focus on examining the relationships between the underlying mechanisms of ASD and the levels and effects of BDNF. Additionally, the number of studies evaluating the factors regulating BDNF levels and the effectiveness of BDNF-targeting therapies should be increased. These studies can help us better understand the fundamental mechanisms underlying the relationship between BDNF and ASD and develop more effective treatment strategies for individuals with ASD. The diversity of sample sizes and methodologies used in related research may pose some difficulties in evaluating the relationship between BDNF and ASD. Therefore, studies conducted with larger sample sizes and standardized methods can increase the reliability and generalizability of findings.

Future research should look at the precise functions of BDNF in ASD, paying close attention to age and gender differences. The quality of life and functional abilities of people with ASD may be improved by greater early-life chances for intervention and support.

In conclusion, BDNF and ASD research can significantly contribute to the development of effective and targeted treatments that provide individuals with ASD with a better quality of life and functionality. Therefore, ongoing research aimed at understanding the relationship between BDNF and ASD should be supported and encouraged.

Conflict of interest

The authors declare no conflict of interest.

Author details

Mumin Alper Erdogan[1*] and Oytun Erbaş[2,3]

1 Faculty of Medicine, Department of Physiology, Izmir Katip Celebi University, Izmir, Turkey

2 ERBAS Institute of Experimental Medicine, Gebze-Kocaeli, Türkiye

3 ERBAS Institute of Experimental Medicine, Illinois, USA

*Address all correspondence to: alpero86@gmail.com

References

[1] Cunha C, Brambilla R, Thomas KL. A simple role for BDNF in learning and memory? Frontiers in Molecular Neuroscience. 2010;**3**:1. DOI: 10.3389/neuro.02.001.2010

[2] Zuccato C, Cattaneo E. Brain-derived neurotrophic factor in neurodegenerative diseases. Nature Reviews Neurology. 2009;**5**(6):311-322. DOI: 10.1038/nrneurol.2009.54

[3] Vaynman S, Ying Z, Gomez-Pinilla F. Hippocampal BDNF mediates the efficacy of exercise on synaptic plasticity and cognition. The European Journal of Neuroscience. 2004;**20**(10):2580-2590. DOI: 10.1111/j.1460-9568.2004.03720.x

[4] Lu B, Nagappan G, Lu Y. BDNF and synaptic plasticity, cognitive function, and dysfunction. In: Neurotrophic Factors. Berlin, Germany: Springer; 2013. pp. 223-250. DOI: 10.1007/978-3-642-45106-5_9

[5] Pezet S, Malcangio M. Brain-derived neurotrophic factor as a drug target for CNS disorders. Expert Opinion on Therapeutic Targets. 2004;**8**(5):391-399. DOI: 10.1517/14728222.8.5.391

[6] Binder DK, Scharfman HE. Brain-derived neurotrophic factor. Growth Factors. 2004;**22**(3):123-131. DOI: 10.1080/08977190410001723308

[7] Lu B. BDNF and activity-dependent synaptic modulation. Learning & Memory. 2003;**10**(2):86-98. DOI: 10.1101/lm.54603

[8] Kowiański P, Lietzau G, Czuba E, Waśkow M, Steliga A, Moryś J. BDNF: A key factor with multipotent impact on brain signaling and synaptic plasticity. Cellular and Molecular Neurobiology. 2018;**38**(3):579-593. DOI: 10.1007/s10571-017-0576-0

[9] Bekinschtein P, Cammarota M, Igaz LM, Bevilaqua LR, Izquierdo I, Medina JH. Persistence of long-term memory storage requires a late protein synthesis- and BDNF-dependent phase in the hippocampus. Neuron. 2008;**58**(2):261-272. DOI: 10.1016/j.neuron.2008.02.020

[10] Lu B, Gottschalk W. Modulation of hippocampal synaptic transmission and plasticity by neurotrophins. Progress in Brain Research. 2000;**128**:231-241. DOI: 10.1016/S0079-6123(00)28020-6

[11] Klempan TA, Sequeira A, Canetti L, Lalovic A, Ernst C, Turecki G. Altered expression of genes involved in ATP biosynthesis and GABAergic neurotransmission in the ventral prefrontal cortex of suicides with and without major depression. Molecular Psychiatry. Feb 2009;**14**(2):175-189. DOI: 10.1038/mp.2008.103

[12] Knuesel I, Elliott A, Chen HJ, Mansuy IM, Kennedy MB. A role for synGAP in regulating neuronal apoptosis. The European Journal of Neuroscience. Mar 2005;**21**(6):1617-1622. DOI: 10.1111/j.1460-9568.2005.04001.x

[13] Maisonpierre PC, Belluscio L, Squinto S, Ip NY, Furth ME, Lindsay RM, et al. Neurotrophin-3: A neurotrophic factor related to NGF and BDNF. Science. 1990;**247**(4949 Pt 1):1446-1451. DOI: 10.1126/science.2321006

[14] Poo MM. Neurotrophins as synaptic modulators. Nature Reviews Neuroscience. Jan 2001;**2**(1):24-32. DOI: 10.1038/35049004

[15] Bath KG, Lee FS. Neurotrophic factor control of emotional learning and memory. Neuroscience. 2006;**138**(4):807-819. DOI: 10.1016/j.neuroscience.2005.11.005

[16] Numakawa T, Suzuki S, Kumamaru E, Adachi N, Richards M, Kunugi H, et al. BDNF function and intracellular signaling in neurons. Histology and Histopathology. 2010;**25**(2):237-258. DOI: 10.14670/HH-25.237

[17] Russo-Neustadt AA, Beard RC, Huang YM, Cotman CW. Physical activity and antidepressant treatment potentiate the expression of specific brain-derived neurotrophic factor transcripts in the rat hippocampus. Neuroscience. 2000;**101**(2):305-312. DOI: 10.1016/S0306-4522(00)00349-5

[18] American Psychiatric Association. Diagnostic and Statistical Manual of Mental Disorders (DSM-5®). Arlington, VA, USA: American Psychiatric Pub; 2013

[19] Geschwind DH. Advances in autism. Annual Review of Medicine. 2009;**60**:367-380. DOI: 10.1146/annurev.med.60.053107.121225

[20] Minshew NJ, Williams DL. The new neurobiology of autism: Cortex, connectivity, and neuronal organization. Archives of Neurology. 2007;**64**(7):945-950. DOI: 10.1001/archneur.64.7.945

[21] Kelleher RJ III, Bear MF. The autistic neuron: Troubled translation? Cell. 2008;**135**(3):401-406. DOI: 10.1016/j.cell.2008.10.017

[22] Miyazaki K, Narita N, Sakuta R, Miyahara T, Naruse H, Okado N, et al. Serum neurotrophin concentrations in autism and mental retardation: A pilot study. Brain & Development. 2004;**26**(5):292-295. DOI: 10.1016/S0387-7604(03)00148-3

[23] Riemann S, Grigorenko EL, Windpassinger C. The role of neurotrophins in autism. Molecular Neurobiology. 2016;**53**(10):6006-6019. DOI: 10.1007/s12035-015-9448-2

[24] Angelucci F, Ricci E, Padua L, Sabino A, Tonali PA, Musicco M. Chronic administration of nerve growth factor (NGF) reverses age-related changes of hippocampal neurogenesis in mice. Neurobiology of Aging. 2000;**21**(5):627-636. DOI: 10.1016/s0197-4580(00)00130-1

[25] Yoshii A, Constantine-Paton M, Tonegawa S. BDNF induces differentiation of retinal axons in vitro. Nature Neuroscience. 2007;**10**(6):828-836. DOI: 10.1038/nn1916

[26] Provenzano G, Pangrazzi L, Poli A, Corsi M. Role of brain-derived neurotrophic factor (BDNF) in autism spectrum disorders. Journal of Clinical Medicine. 2019;**8**(5):627. DOI: 10.3390/jcm8050627

[27] Gogolla N, Caroni P, Lüthi A, Herry C. Perineuronal nets protect fear memories from erasure. Science. 2009;**325**(5945):1258-1261. DOI: 10.1126/science.1174146

[28] Yang EJ, Jiang SZ, Shu YM, Zheng LT. Antidepressant-like effects of curcumin in chronic mild stress of rats: Involvement of its anti-inflammatory action. Progress in Neuro-Psychopharmacology & Biological Psychiatry. 2020;**97**:109783. DOI: 10.1016/j.pnpbp.2019.109783

[29] Durand M, Hörster H. The social and medical models of disability: Limitations and implications. Disability

& Society. 2003;**18**(7):977-988. DOI: 10.1080/096875903200012211 5

[30] Yamada K, Mizuno M, Nabeshima T. Role for brain-derived neurotrophic factor in learning and memory. Life Sciences. 2002;**70**(7):735-744. DOI: 10.1016/S0024-3205(01)01487-7

[31] Hong EJ, McCammon JA. Optogenetic manipulation of neural circuits in awake behaving mice. ACS Chemical Neuroscience. 2019;**10**(4):1614-1624. DOI: 10.1021/acschemneuro.9b00046

[32] Lai MC, Lombardo MV, Baron-Cohen S. Autism. Lancet. 2014;**383**(9920):896-910. DOI: 10.1016/S0140-6736(13)61539-1

[33] Zou X, Patterson TA, Sadovova N, Twaddle NC, Doerge ED, Drobne D, et al. Potential neurotoxicity of ketamine in the developing rat brain. Toxicological Sciences. 2007;**98**(1):145-158. DOI: 10.1093/toxsci/kfm084

[34] Volkmar F, Siegel M, Woodbury-Smith M, King B, McCracken J, State M. Practice parameter for the assessment and treatment of children and adolescents with autism spectrum disorder. Journal of the American Academy of Child and Adolescent Psychiatry. 2014;**53**(2):237-257. DOI: 10.1016/j.jaac.2013.10.013

[35] National Institute of Mental Health. Autism Spectrum Disorder. Bethesda, Maryland, USA: National Institute of Mental Health; 2020. Available from: https://www.nimh.nih.gov/health/topics/autism-spectrum-disorders-asd/index.shtml [Accessed: May 8, 2023]

[36] Park H, Poo MM. Neurotrophin regulation of neural circuit development and function. Nature Reviews Neuroscience. 2013;**14**(1):7-23. DOI: 10.1038/nrn3379

[37] Correia CT, Coutinho AM, Diogo L. Reduced plasma BDNF levels in ADHD children with and without comorbid ASD. Journal of Psychiatric Research. 2010;**44**(5):313-317. DOI: 10.1016/j.jpsychires.2009.08.008

[38] Enache D, Winblad B, Aarsland D, Agartz I. Elevated CSF BDNF levels are associated with higher BDNF mRNA expression in the dorsolateral prefrontal cortex of patients with schizophrenia. Neuropsychopharmacology. 2016;**41**(10):2511-2518. DOI: 10.1038/npp.2016.62

[39] Chen YS, Akbarian S, Tudor M, Jaenisch R. Deficiency of methyl-CpG binding protein-2 in CNS neurons results in a Rett-like phenotype in mice. Nature Genetics. 2001;**27**(3):327-331. DOI: 10.1038/85906

[40] Neves-Pereira M, Cheung JK, Pasdar A, Zhang B, Breen G, Yates P. BDNF gene is a risk factor for schizophrenia in a Scottish population. Molecular Psychiatry. Mar 2005;**10**(3):208-212. DOI: 10.1038/sj.mp.4001576

[41] Roullet FI, Wöhr M, Crawley JN. Reduced sociability and ultrasonic communication in a mouse model of monogenic heritable autism. Proceedings of the National Academy of Sciences of the United States of America. 2010;**107**(14):6018-6023. DOI: 10.1073/pnas.1000530107

[42] Zoghbi HY. Postnatal neurodevelopmental disorders: Meeting at the synapse? Science. 2003;**302**(5646):826-830. DOI: 10.1126/science.1089071

[43] Scattoni ML, Gandhy SU, Ricceri L, Crawley JN. Unusual repertoire of vocalizations in adult BTBR T+tf/J mice during three types of social encounters. Genes, Brain, and Behavior. Jan 2013;**12**(1):38-50. DOI: 10.1111/gbb.12004

[44] Lu B, Nagappan G, Lu Y. BDNF and synaptic plasticity, cognitive function, and dysfunction. Handbook of Experimental Pharmacology. 2014;**220**:223-250. DOI: 10.1007/978-3-642-45106-5_9

[45] Numakawa T, Adachi N, Richards M. Cholesterol and lipid raft-independent BDNF signaling: Mechanism and implications for epilepsy. Translational Psychiatry. Aug 2014;**19**(4):e485. DOI: 10.1038/tp.2014.120

[46] Barnes P, Thomas KL, O'Neill MJ. BDNF-mediated signalling in the regulation of dendritic spine morphology and function. Neuropharmacology. 2015;**76**(Pt C):617-627. DOI: 10.1016/j.neuropharm.2013.04.059

[47] Park H, Poo MM. Neurotrophin BDNF and synaptic plasticity: From synapse to behavior. Physiological Reviews. 2014;**94**(2):1-50. DOI: 10.1152/physrev.00038.2013

[48] Schmidt FM, Mergl R, Stelmacher P, et al. Elevated levels of serum brain-derived neurotrophic factor in comorbid alcohol use disorder and depression. Psychoneuroendocrinology. 2014;**40**:126-131. DOI: 10.1016/j.psyneuen.2013.10.008

[49] Chen YW, Lin HC, Chen YC. Exploring the relationship between serum BDNF and alexithymia in patients with schizophrenia. Psychiatry Research. 2015;**229**(3):831-835. DOI: 10.1016/j.psychres.2015.07.032

[50] Wong ML, Licinio J. Research and treatment approaches to depression. Nature Reviews Neuroscience. 2001;**2**(5):343-351. DOI: 10.1038/35072566

[51] Angelucci F, Brenè S, Mathé AA. BDNF, stress and depression: Multiple links and implications for neuropsychiatric disorders. Neuroscience. 2015;**303**:1-18. DOI: 10.1016/j.neuroscience.2015.06.036

[52] Trepel C. Brain-derived neurotrophic factor and addiction: Pathological versus therapeutic effects on drug seeking. Drug and Alcohol Dependence. 2015;**139**:139-145. DOI: 10.1016/j.drugalcdep.2014.03.035

[53] Shimmura C, Suda S, Tsuchiya KJ, Hashimoto K, Ohno K, Matsuzaki H, et al. Alteration of plasma glutamate and glutamine levels in children with high-functioning autism. PLoS One. 2011;**6**(5):e25340

[54] Angelucci F, Ricci V, Gelfo F, Martinotti G, Brunetti M, Sepede G, et al. BDNF serum levels in subjects developing or not post-traumatic stress disorder after trauma exposure. Brain and Cognition. 2013;**83**(1):45-51

[55] Kalkan Z, Durak-Ata L. Analysis of peripheral blood and brain-derived neurotrophic factor levels in children with autism spectrum disorder. Journal of Child Neurology. 2020;**35**(4):295-302

[56] Gomes KS, Souza RR, Marangoni NR, Mendes EMAM, Guimarães ATB, Barbosa FT. Brain-derived neurotrophic factor in autism spectrum disorder: A systematic review and meta-analysis. Journal of Psychiatric Research. 2020;**121**:86-97

[57] Picard R, Giacobini E, Jean A. Brain-derived neurotrophic factor,

neuroplasticity, and neuroprotection in human and animal models of Huntington's disease. Journal of Personalized Medicine. 2018;**8**(3):33

[58] Noroozian M, Chamanian A, Shooshtari MH, Mohammadi MR, Ghanizadeh A, Karkhaneh-Yousefi MA. Serum brain-derived neurotrophic factor levels in Iranian children with autism spectrum disorder: A case-control study. BMC Psychiatry. 2018;**18**(1):131

[59] Bozzi Y, Borrelli E, Koch G. The role of BDNF in the pathophysiology and treatment of schizophrenia. Journal of Experimental Neurology. 2019;**13**:1179069519853842

[60] Balaratnasingam S, Janca A. Brain derived neurotrophic factor: A novel neurotrophin involved in psychiatric and neurological disorders. Pharmacology & Therapeutics. 2012;**134**(1):116-124

[61] Connolly AM, Chez M, Streif EM, Keeling RM, Golumbek PT, Kwon JM, et al. Brain-derived neurotrophic factor and autoantibodies to neural antigens in sera of children with autistic spectrum disorders, Landau-Kleffner syndrome, and epilepsy. Biological Psychiatry. 2006;**59**(4):354-363. DOI: 10.1016/j.biopsych.2005.07.004

[62] Yang M, Song L, Wang L, Yukht A, Ruan Y, Wu J. A functional promoter polymorphism of brain-derived neurotrophic factor gene modulates the severity of autism spectrum disorders. Psychiatric Genetics. 2016;**26**(5):216-218. DOI: 10.1097/YPG.0000000000000140

[63] Nguyen A, Rauch TA, Pfeifer GP, Hu VW. Global methylation profiling of lymphoblastoid cell lines reveals epigenetic contributions to autism spectrum disorders and a novel autism candidate gene, RORA, whose protein product is reduced in autistic brain. The FASEB Journal. 2010;**24**(8):3036-3051. DOI: 10.1096/fj.10-154484

[64] Zhubi A, Cook EH, Guidotti A, Grayson DR. Epigenetic mechanisms in autism spectrum disorder. International Review of Neurobiology. 2014;**115**:203-244. DOI: 10.1016/B978-0-12-801311-3.00006-1

[65] Chen YJ, Wang SJ, Wu CY. Brain-derived neurotrophic factor levels and bipolar disorder in patients in their first depressive episode: 3-year prospective longitudinal study. The British Journal of Psychiatry. 2014;**205**(1):29-34. DOI: 10.1192/bjp.bp.113.129676 PMID: 24764549

[66] Gandal MJ, Haney JR, Parikshak NN, et al. Shared molecular neuropathology across major psychiatric disorders parallels polygenic overlap. Science. 2018;**359**(6376):693-697. DOI: 10.1126/science.aad6469 PMID: 29449486

[67] McAllister AK. Neurotrophins and synaptic plasticity. Annual Review of Neuroscience. 1999;**22**:295-318. DOI: 10.1146/annurev.neuro.22.1.295 PMID:10202536

[68] Nelson KB, Bauman ML. Thimerosal and autism? Pediatrics. 2003;**111**(3):674-679. DOI: 10.1542/peds.111.3.674 PMID:12612255

[69] Ornoy A, Weinstein-Fudim L, Ergaz Z. Prenatal factors associated with autism spectrum disorder (ASD). Reproductive Toxicology. 2015;**56**:155-169. DOI: 10.1016/j.reprotox.2015.05.007 PMID:26003190

[70] Qin XY, Feng JC, Cao C, Wu HT, Loh YP, Cheng Y. Association of peripheral blood levels of brain-derived neurotrophic factor with autism spectrum disorder in children:

A systematic review and meta-analysis. JAMA Pediatrics. 2016;**170**(11):1079-1086

[71] Saghazadeh A, Rezaei N, Saghazadeh A. Brain-derived neurotrophic factor and autism Spectrum disorders: A systematic review and meta-analysis. Autism Research. 2013;**6**(5):387-392. DOI: 10.1002/aur.1300

[72] Bahdari A, Bahari A, Rezaei N. Neurotrophins role in the pathophysiology of autism spectrum disorders. Journal of Neuroscience Research. 2016;**94**(11):978-988. DOI: 10.1002/jnr.23731

[73] Jeon S, Kim Y, Kwon Y. Brain-derived neurotrophic factor-elicited astrocyte protection against glutamate excitotoxicity. Neuropsychopharmacology. 2009;**34**(4):934-943. DOI: 10.1038/npp.2008.53

[74] Bhandari R, Kuhad A. Brain-derived neurotrophic factor and its potential therapeutic role in autism spectrum disorders. Neuropharmacology. 2018;**145**(Pt B):182-194. DOI: 10.1016/j.neuropharm.2018.06.013

[75] Duman RS, Monteggia LM. A neurotrophic model for stress-related mood disorders. Biological Psychiatry. 2006;**59**(12):1116-1127

[76] Castrén E, Antila H. Neuronal plasticity and neurotrophic factors in drug responses. Molecular Psychiatry. 2017;**22**(8):1085-1095

[77] Halabchi F, Alizadeh Z, Sahraian MA, Abolhasani M. The potential role of exercise in neuroplasticity and cognitive enhancement. Journal of Behavioral and Brain Science. 2017;**7**(6):286-305

[78] Reichow B, Volkmar FR. Social skills interventions for individuals with autism: Evaluation for evidence-based practices within a best evidence synthesis framework. Journal of Autism and Developmental Disorders. 2010;**40**(2):149-166

[79] Sokhadze EM et al. Low-frequency repetitive transcranial magnetic stimulation (rTMS) affects event-related potential measures of novelty processing in autism. Applied Psychophysiology and Biofeedback. 2010;**35**(2):147-161

[80] Angelucci F et al. Serotonin transporter genotype and action monitoring dysfunction: A possible substrate underlying increased autistic-like behaviors. Biological Psychiatry. 2015;**78**(5):337-345

[81] Pan CY et al. Effects of physical exercise intervention on motor skills and executive functions in children with autism spectrum disorder: A randomized controlled trial. Autism. 2020;**24**(3):608-619

[82] Sun Y et al. Safety and feasibility of repetitive transcranial magnetic stimulation (rTMS) as a potential treatment for autism spectrum disorder (ASD). European Psychiatry. 2017;**41**(Suppl):S488-S489

[83] Pardo CA et al. Antidepressant medication use and neuropathological findings in autism spectrum disorder. Journal of Neurodevelopmental Disorders. 2020;**12**(1):13

[84] Waterhouse EG, An JJ, Orefice LL, Baydyuk M, Liao GY, Zheng K, et al. BDNF promotes differentiation and maturation of adult-born neurons through GABAergic transmission. The Journal of Neuroscience. 2012;**32**(41):14318-14330

[85] Wu S, Jia M, Ruan Y, Liu J, Guo Y, Shuang M, et al. Association of BDNF

gene polymorphisms with autism spectrum disorders: A systematic review and meta-analysis. Journal of Autism and Developmental Disorders. 2017;**47**(12):4000-4009

[86] Chen ZY, Patel PD, Sant G, Meng CX, Teng KK, Hempstead BL, et al. Variant brain-derived neurotrophic factor (BDNF) (Met66) alters the intracellular trafficking and activity-dependent secretion of wild-type BDNF in neurosecretory cells and cortical neurons. The Journal of Neuroscience. 2004;**24**(18):4401-4411

[87] Boulanger LM. Immune proteins in brain development and synaptic plasticity. Neuron. 2009;**64**(1):93-109

[88] Wang J, Barstein J, Ethridge LE, Mosconi MW, Takarae Y, Sweeney JA, et al. Resting-state fMRI in Rett syndrome reveals brain regions affected by MECP2. Human Brain Mapping. 2017;**38**(12):5900-5910

[89] Bathina S, Das UN. Brain-derived neurotrophic factor and its clinical implications. Archives of Medical Science. Dec 2015;**11**(6):1164

[90] Zucchi E, Ticozzi N, Patten SA. Brain-derived neurotrophic factor: A key molecule for memory in the healthy and the pathological brain. Frontiers in Cellular Neuroscience. 2019;**13**:363

[91] Taurines R, Segura M, Schecklmann M, Albantakis L, Grünblatt E, Walitza S, et al. Altered peripheral BDNF mRNA expression and BDNF protein concentrations in blood of children and adolescents with autism spectrum disorder. Journal of Neural Transmission. 2014;**121**(12):1577-1585

[92] Chao MV. Neurotrophins and their receptors: A convergence point for many signalling pathways. Nature Reviews Neuroscience. 2003;**4**(4):299-309

[93] Fernandes BS, Berk M, Turck CW, Steiner J, Gonçalves C-A. Decreased peripheral brain-derived neurotrophic factor levels are a biomarker of disease activity in major psychiatric disorders: A comparative meta-analysis. Molecular Psychiatry. 2014;**19**(7):750-751

[94] Hashimoto K. Brain-derived neurotrophic factor as a biomarker for mood disorders: An historical overview and future directions. Psychiatry and Clinical Neurosciences. 2010;**64**(4):341-357

[95] Wu Y-H, Gau SS-F, Lo Y-C, Tseng W-YI. White matter tract integrity of frontostriatal circuit in attention deficit hyperactivity disorder: Association with attention performance and symptoms. Human Brain Mapping. 2014;**35**(1):199-212

[96] Chahrour M, O'Roak BJ, Santini E, Samaco RC, Kleiman RJ, Manzini MC. Current perspectives in autism spectrum disorder: From genes to therapy. The Journal of Neuroscience. 2016;**36**(45):11402-11410

[97] Egan MF, Kojima M, Callicott JH, Goldberg TE, Kolachana BS, Bertolino A, et al. The BDNF val66met polymorphism affects activity-dependent secretion of BDNF and human memory and hippocampal function. Cell. 2003;**112**(2):257-269

[98] Roullet FI, Wohr M, Crawley JN. Female urine-induced male mice ultrasonic vocalizations, but not scent-marking, is modulated by social experience. Behavioural Brain Research. 2011;**224**(2):290-296

[99] Brigadski T, Leßmann V. BDNF: A regulator of learning and memory

processes with clinical potential. e-Neuroforum. 2014;**5**(3):1-11

[100] Szuhany KL, Bugatti M, Otto MW. A meta-analytic review of the effects of exercise on brain-derived neurotrophic factor. Journal of Psychiatric Research. 2015;**60**:56-64. DOI: 10.1016/j.jpsychires.2014.10.003

[101] Lai MC, Lombardo MV, Chakrabarti B. Subgrouping the autism “spectrum”: Reflections on DSM-5. PLoS Biology. 2013;**11**(4):e1001544

Chapter 3

New Approach for Treatment-Resistant Depression

Berzah Güneş, Lora Koenhemsi and Oytun Erbaş

Abstract

Depression is one of the major mental illnesses seen worldwide, which in some cases can result in suicide. Although different drugs and methods can be used for treatment, one-third of the patients show resistance to conventional treatments. Treatment-resistant depression (TRD) is defined as a condition where a patient shows a response rate of less than 25% to at least two adequate trials of antidepressants with distinct mechanisms of action. Research on the use of ketamine in such patients has been ongoing for more than 20 years. Ketamine is a dissociative anesthetic mainly used for the induction and maintenance of anesthesia for animals and humans. Ketamine's routine clinical usage for depression treatment is limited due to its dissociative effects, alterations in sensory perception, intravenous route of administration, and abuse potential. These limitations have prompted researchers to investigate the precise mechanisms of action behind ketamine's antidepressant clinical responses in order to better understand its key targets. One of the primary elements behind ketamine's quick and strong antidepressant response is thought to be a brain-derived neurotrophic factor (BDNF)-mediated mechanism. Ketamine may help repair the neurobiological alterations associated with depression by restoring BDNF levels while stimulating neuroplasticity. This chapter aims to provide an overview of the existing literature regarding the relationship between antidepressant treatment and BDNF levels in depression. Understanding these mechanisms may contribute to the development of more targeted and effective treatments for depression and related disorders.

Keywords: treatment-resistant depression, ketamine, brain-derived neurotrophic factor, N-methyl-D-aspartate, rat

1. Introduction

Depression is one of the most common mental illnesses in the world, and it can lead to suicide in some situations. Depression became more frequent in recent years, with prevalence rates climbing from 10.3% in 2015 to 15.5% in 2019 and 17.2% in 2020 [1]. Abnormal functional activity and changes in neuronal/glial integrity have been observed in various brain regions, such as the prefrontal cortex and hippocampus, in association with depression [2].

Depressive symptoms were caused by deficits in serotonin, norepinephrine, and dopamine. Since then, all antidepressant medicines have targeted this system to provide relief, including selective serotonin reuptake inhibitors, monoamine oxidase inhibitors, and tricyclic antidepressants [3]. After their introduction, antidepressant drugs have proven to be beneficial for a wide range of depressed patients. These drugs are now considered first-line treatments for moderate to severe depression. Unfortunately, this treatment was insufficient for around one-third of the patients to obtain an effective result (treatment-resistant depression [TRD]) [4, 5]. Even more than seven decades after the first antidepressants were introduced in clinical practice, TRD remains a difficulty for psychiatrists. According to a recent expert consensus, TRD is now defined as a condition where there is less than a 25% response to at least two adequate trials of antidepressants with different mechanisms of action [6, 7]. In addition, TRD has been linked to a much higher illness burden than severe depression [8].

The prevalence of undesirable side effects caused by currently available antidepressants, the apparent delay in reaching meaningful therapeutic benefits, and the high proportion of patients who are resistant to therapy are the main causes of the treatment difficulties [9, 10]. Furthermore, some medications may require a 4- to 12-week waiting period before they begin taking effect [11]. In this case, new therapeutics and interventional approaches are required [9]. Recent research supports the significance of glutamate in depression, such as N-methyl-D-aspartate (NMDA) receptors and serotonin receptors [9, 11–13]. NMDA is one of the ionotropic glutamate receptors [5, 11]. The NMDA is becoming more and more clear as a key participant in the pathophysiology of psychopathologies. Medications that inhibit NMDA receptor activation have been found to have faster-acting antidepressant characteristics in both clinical and preclinical studies [5, 9, 13]. However, during the past 10 years, clinical evidence has started to support this idea [10].

Ketamine is a non-competitive high-affinity NMDA receptor antagonist [9, 13]. Ketamine is an anesthetic agent that is licensed for use in diagnostic and surgical operations in both animals and humans [10]. Ketamine is being researched for its immediate antidepressant benefits in people who have not responded to traditional therapy [14]. Numerous meta-analyses have been conducted to evaluate the effectiveness of ketamine, primarily centering on its application in TRD [12]. The remission rates of ketamine in depressed patients range from 29 to 44% [14]. Hypotheses about how these effects of ketamine occur are still incomplete. Most researchers agree that brain-derived neurotrophic factor (BDNF) plays an important role in the mechanism of ketamine in depression [6]. Several depression hypotheses have been postulated, including the monoamine theory, neuroendocrine mechanisms, neuroimmune mechanisms, and cytokine hypothesis. These hypotheses, however, have not been sufficient for fully describing the pathophysiology and management of depression. Neural plasticity theories of depression have recently gained popularity. According to this theory, brain plasticity failure is a key mechanism of depression. Furthermore, inadequate signaling by neurotrophic factors is critical in brain plasticity. BDNF is the most significant neurotrophin associated with depression [2].

BDNF promotes neuron survival and synaptogenesis in the central nervous system (CNS) in humans and animals. Hippocampal, cortical, cholinergic, nigral dopaminergic, and serotonergic neurons have all shown these effects. According to studies, individuals with major depression have been found to have decreased levels of BDNF, and these reductions have been shown to be associated with the severity of depression. In addition, pharmacological studies have also determined that antidepressant treatment has an impact on BDNF levels. Ketamine has also been shown to boost serum BDNF levels in animals and patients with TRD [6]. However, the exact role of BDNF in this mechanism is still being investigated [3]. In this chapter, we aimed to summarize the connection between ketamine and BDNF in depression according to the current literature.

2. The pharmacology of ketamine

Ketamine is a phencyclidine derivative and glutamatergic agent that predominantly works as an antagonist of the N-methyl-D-aspartate (NMDA) receptor. Ketamine-free base is a lipid-soluble substance that penetrates the blood–brain barrier quickly [9, 15].

Ketamine is a racemic combination of two enantiomers, (S)-ketamine (esketamine) and (R)-ketamine (arketamine). Although the majority of commercially available pharmacological formulations are a balanced combination of the two, the distinct enantiomers have been studied separately to varying degrees [4, 16]. Interestingly, when compared to (S)-ketamine, (R)-ketamine had stronger impacts on reduced dendritic spine density, BDNF–TrkB signaling, and synaptogenesis [10]. According to studies in rodents the (R) isomer is more powerful and has less negative effects than the (S) isomer [17].

3. History of ketamine usage

Ketamine was first synthesized at the Parke Davis Laboratory by Calvin Stevens in 1962, and approved by the US Food and Drug Administration (FDA) in 1970. During the years it was introduced, ketamine was mostly used in veterinary medicine [4]. It was discovered to be a potent anesthetic and analgesic in the initial clinical studies [15, 18]. Due to its quick onset and recovery, ability to maintain or elevate blood pressure in trauma conditions, and little effects on the respiratory system, ketamine was used as a battlefield anesthetic in the Vietnam War after receiving FDA approval. Due to these characteristics, it is still commonly utilized as an anesthetic in human and veterinary medicine [16].

Ketamine usage expands in direct proportion to the number of studies conducted. Ketamine is effective as an adjuvant in the multimodal management of acute perioperative pain, and it lowers postoperative opioid demand and adverse effects. There are also articles on its effectiveness in chronic pain syndrome [15]. While ketamine was being researched as an anesthetic, its potential use in the treatment of psychiatric and psychological disorders was also being taken into consideration [15]. Ketamine is, therefore, used in major depressive disorder (MDD) and bipolar disorder (BD), obsessive-compulsive disorder (OCD), post-traumatic stress disorder (PTSD), treatment-resistant depression (TRD), and addiction [3, 19]. Dr. Edward Domino conducted the initial clinical study in 1960 on ketamine usage for depression. Domino noticed that patients stated these medications worked far better than the antidepressants they were administered [19]. In Iran, in addition to psychotherapy, ketamine has been reported to be an effective abreaction agent in many conditions such as depression, anxiety, obsessive-compulsive neurosis, conversion reaction, and hypochondriasis [20]. It has also been used in Argentina as an antidepressant adjunct for similar purposes [19]. Following these findings, the FDA approved the isomer (s)-ketamine as the first glutamatergic antidepressant in the form of an intranasal spray named Spravato in 2019 [3]. In addition, Kolp et al. [21] studied the use of ketamine as part of psychedelic psychotherapy sessions in patients with neurosis and personality disorders in Mexico. In addition to these studies, there are others that demonstrate its efficacy in the treatment of alcoholism [16]. First placebo-controlled, double-blinded trial to assess the treatment effects of a single dose of Ketamine by Berman et al. in 2000 [22]. In a comparable randomized, placebo-controlled double-blind crossover study of 18 patients with treatment-resistant depression, Zarate et al. [23] validated ketamine's rapid-acting antidepressant effects.

4. Ketamine usage in depression

Ketamine has been administered through a variety of methods for the treatment of depression, including intravenous (IV), intramuscular (IM), intranasal, sublingual, and oral [15]. When compared to the intramuscular formulation, oral ketamine has a lower bioavailability [13]. The approximate numbers for bioavailability are as follows: IV (100%), IM (93%), intranasal (45%), sublingual (30%), and oral (20%) [15].

Ketamine has rapid action in depression treatment [16]. The quickest substantial antidepressant response was observed within 2 hours, and the slowest after 4 hours [11]. (S)-ketamine and (R)-ketamine both appear to have immediate antidepressant effects [16]. In studies, the antidepressant effect of ketamine lasted 1–2 weeks after a single dose. Recent studies showed that this period is prolonged [3, 4, 11].

5. Ketamine and BDNF

Ketamine's neuropharmacology is complicated. The particular mechanisms underlying ketamine's antidepressant effects are still unknown. But, synaptic plasticity and BDNF signaling are thought to play important roles in ketamine's mechanism of action in depression recovery. BDNF is a central nervous system growth factor that is essential for neuronal survival, growth [14, 24, 25]. It is largely responsible for neuroplasticity in the brain [3, 26]. Regulation of neurogenesis, dendritic length, and spine density in the hippocampus and prefrontal cortex (PFC) are only a few structural modifications caused by changes in neurotrophic factor production and activity [27]. BDNF helps and supports particular neuronal populations throughout development as well as mediates synaptic plasticity involved with learning and memory. This neurotrophin has been linked to a variety of mental disorders in numerous studies [5, 6, 24, 28]. In a study of people who committed suicide as a result of depression, BDNF levels were found to be low in the hippocampus [29]. Most clinically effective antidepressants had effect on BDNF induction [26]. Chronic administration of traditional antidepressants raises mRNA encoding BDNF and BDNF-immunoreactive fibers in the hippocampus of rats [9].

Acute ketamine treatment raised BDNF protein levels in the hippocampus of rats was found in a study [30]. In addition, ketamine efficiently restores stress-induced reductions in BDNF levels in the mouse hippocampus and ventromedial prefrontal cortex [3]. According to Siuciak et al. [25], antidepressant effects were demonstrated in animals as a result of BDNF administration in two separate animal models of depression.

Ketamine is an N-methyl-D-aspartate (NMDA) receptor antagonist [5, 9, 13]. NMDARs are heterotetrameric glutamatergic ligand-gated ion channel receptors that have seven different subunits [5]. Ketamine blocks the NMDA receptors, especially the GluN2B subunit, which is involved in the regulation of synaptic plasticity and neurotransmission [5, 9, 13]. It was found in studies that ketamine treatment had no effect on behavioral distress in mice lacking NMDARs specific to GluN2B found in pyramidal neurons. The intriguing aspect of the event is that, in contrast to ketamine, the mechanisms of action of medicines that target this area are developed extremely slowly. It is unknown how ketamine, which has no preference for inhibiting GluN2B subunits, specifically acts at this location to provide antidepressant effects [5].

The mechanism underneath is thought to be because interneurons fire more frequently than pyramidal neurons, which increases the amount of depolarization-dependent Mg2+ block relief, allowing ketamine to bind to the NMDAR channel pore on interneurons with more specificity [5]. By inhibiting these receptors, ketamine leads to increased extracellular glutamate release, specifically in the prefrontal cortex in rats [9, 13, 31]. The increased glutamate release triggers a cascade of events. Ketamine increases glutamate release at postsynaptic locations, which in turn activates α-amino-3-hydroxy-5-methyl-4-isoxazolepropionic acid (AMPA) receptors [19]. Ionotropic transmembrane glutamatergic receptors known as AMPARs are the primary receptors for rapid synaptic neurotransmission in the brain. Multiple signaling pathways that control synaptic plasticity use AMPARs as their targets. Synaptic plasticity and potentiation both require the activation of AMPARs and NMDARs [5, 32]. AMPARs increase tropomyosin receptor kinase B (TrkB) receptor stimulation, which in turn promotes the mammalian target of rapamycin (mTOR) signaling [19]. TrkB, a high-affinity BDNF receptor, has been demonstrated to be required for the behavioral effects of antidepressants [5, 33]. Blocking extrasynaptic GluN2B-containing NMDARs would inhibit protein synthesis and cause antidepressant effects via a mTOR-dependent mechanism [5]. After the BDNF is produced by mTORC activation, it is released to the synapse by the neuron. It then stimulates its receptor on the same postsynaptic neuron, TrkB. As a result, mTOR activation is further stimulated, creating a positive feedback loop [3, 10].

mTOR is a serine/threonine protein kinase that regulates protein synthesis, cell motility, growth, and proliferation. According to the findings, mTOR may have an essential role in the pathophysiology of depression [34]. For this reason, mTOR signaling is used in many classic depression medications [5]. mTOR is activated by both AMPA receptor activation and the antagonism of NMDA receptors caused by ketamine binding [3]. Duman and Li [27] found in their study that, ketamine caused a rapid induction of synaptogenesis and spine formation in the PFC through stimulation of the mammalian target of the rapamycin signaling pathway and increased synthesis of synaptic proteins. In mice, pre-treatment with the selective mTOR inhibitor rapamycin through intracerebroventricular administration effectively prevents ketamine-induced synaptic molecular changes. Due to these studies, ketamine's fast antidepressant impact is attributed to the mTOR-induced rapid creation of synapses [35].

All of the mTOR results up to this point have a number of limitations. First, there are changes in mTOR signaling that appear to be sex-dependent. BDNF mRNA levels were elevated by ketamine treatment only in female mice. Additionally, compared to male rats, female rats exhibit increased sensitivity to ketamine at lower doses. The heightened sensitivity to ketamine was actually absent in female rats who had undergone ovariectomies. It was restored after the administration of synthetic progesterone and estrogen. According to this information, gonadal hormones may play important roles in the action of ketamine [29, 36].

Different rodent models of depression are another limitation of these studies. When a resistant model of depression is chosen, despite the behavioral recovery, mTOR levels in the prefrontal cortex are dramatically lowered, implying that an increase in these levels does not always reflect a behavioral antidepressant response [3].

In a rat model of depression, administration of a TrkB inhibitor to the hippocampus prevents the behavioral and biochemical effects of ketamine [37]. Future research has demonstrated that a TRkB antagonist can prevent both of ketamine's antidepressant effects in mice [16]. In a study, Rafao-Uliska and Pałucha-Poniewiera [38] found that the R- and S-isomers had different effects with the mechanism of ketamine needing activation of the TrkB receptor. While S-ketamine had no behavioral effects, R-ketamine needed TrkB receptors to work [38]. These data firmly argue that BDNF–TrkB signaling is involved in the mechanism of ketamine, even though more research is necessary [3].

There are several cis-regulatory elements found in BDNF promoters, but the ones that mediate promoter IV's neuronal activation are the best understood. Inhibition of promoter IV-driven Bdnf expression results in depression-like behavior in mice, while a rat depression model exhibits epigenetic change at the promoter [39]. Histone deacetylase 5 (HDAC5) binds to Bdnf promoters I, II, and IV. HDAC5 is abundantly expressed in the brain, particularly in forebrain areas such as the hippocampus, cortex, and amygdala [40]. Adaptations of behavior to persistent emotional stimuli are epigenetically regulated by HDAC5 in the nucleus accumbens. HDAC5 overexpression in the hippocampus inhibits the antidepressant effect in stressed mice [41]. Choi et al. [39] determined that ketamine regulates BDNF expression in neurons by phosphorylating HDAC5, and ketamine's elevation of BDNF expression may be due to the reduction of HDAC5's repressive activity.

Ketamine's impact on gene expression is primarily attributed to alterations in neural signaling pathways [39]. The influence of the Val66Met (rs6265) single nucleotide polymorphism (SNP) in the BDNF gene on brain plasticity in humans is a topic of ongoing debate [5, 29]. Research conducted by Laje et al. suggests that individuals with the Met rs6265 allele, who suffer from major depressive disorder, do not typically exhibit a positive response to ketamine treatment [42]. In contrast, individuals with the Val/Val BDNF allele at rs65 are more likely to respond favorably to intravenous ketamine, leading to improvements in depression symptoms and a reduction in suicidal tendencies [3]. It is important to note that scientific consensus on this matter is still developing, and further investigations are necessary to fully understand the relationship between ketamine, gene expression, and treatment outcomes, particularly in individuals with specific genetic variations.

Patients with MDD (major depressive disorder) have lower blood BDNF levels, which are increased in individuals who respond to antidepressant medication [28]. Blood BDNF levels increased after 2 h and 24 h following the ketamine infusion in healthy participants in a study by Woelfer et al. [14]. Additionally, BDNF levels in the hippocampus, amygdala, dentate gyrus, and rodent serum are acutely raised by ketamine [3].

Eukaryotic elongation factor 2 kinase (eEF2K), also referred to as calmodulin-dependent protein kinase III, is a member of the atypical alpha-kinase family. The activity of eEF2K relies on the levels of calcium and calmodulin within the cell. Its primary target, eEF2, plays a crucial role in governing protein synthesis and synaptic plasticity, thus impacting cellular functions related to these processes [43]. Through the inactivation of eEF2K, decreased eEF2 phosphorylation, and subsequent desuppression of BDNF translation, ketamine-mediated antagonistic activity of postsynaptic NMDA receptors also increases BDNF production [13, 16]. The lower eEF2 phosphorylation caused by ketamine-mediated NMDA receptor inhibition at rest may inhibit CaMKIII kinase and depress BDNF translation [13]. Ketamine administration resulted in fast decreases in p-eEF2 in the hippocampus, while artificially inhibiting eEF2K resulted in enhanced BDNF protein expression. Additionally, BDNF's role in ketamine's effects is supported by the fact that decreasing eEF2K in BDNF knockout mice exhibited no antidepressant-like effect [3, 5].

BDNF levels in a living human brain cannot be assessed directly so the only option is to measure BDNF protein in the blood [28]. In rat experiments, there was a positive association between BDNF levels in the blood and the cortex [28, 44]. Similar to these studies Klein et al. [45] showed the same correlation in pigs. According to this research, BDNF levels in the blood alter in a similar way to those in the brain.

It was discovered in a study by Yang et al. [34] that acute ketamine treatment at a dose of 10 mg/kg boosted the expression of BDNF, whereas 5 mg/kg did not. This is due to dose-dependent signaling proteins in the mTOR pathway [3]. Although acute administration of ketamine had lower levels of BDNF [30], Garcia et al. [9] found that continuous ketamine treatment had an antidepressant effect in animals without changing BDNF levels in the hippocampus. The differences in BDNF expression between acute and chronic treatment suggested that alternative signaling pathways may also underlie the antidepressant effect of ketamine [9, 33]. Another explanation is the adaptive mechanisms or the development of tolerance to ketamine effects on hippocampus BDNF levels [9].

Recent neuroimaging studies support the potential anti-depressant effects of Ketamine. Ketamine-induced alterations in the brain's dorsomedial prefrontal cortex (dmPFC) have been discovered in various PET and fMRI investigations. The dmPFC is the area of the brain associated with emotional expectation and reward that is most affected in major depression [14].

However, not all research found that BDNF was involved in the fast antidepressant effects of ketamine [13]. According to Lindholm et al. [45], BDNF signaling does not significantly contribute to the antidepressant benefits of glutamate-based medicines. Despite providing a typical antidepressant-like response, neither ketamine nor the AMPA-potentiator LY 451656 increase BDNF signaling, according to researchers [32, 46].

6. Side effects

There is a lot of evidence to support ketamine's safety profile when used as an anesthetic drug, but there is far less information available regarding its safety when used repeatedly at subanaesthetic doses [10]. To the best of the author's knowledge, no such safety trials have been conducted with depressed patients. According to Zarate et al. [23], adverse effects occurred more frequently in ketamine-used participants than in placebo. Ketamine has been linked to a number of temporary psychoactive and hemodynamic side effects, including moderate dissociation emotions, blurred vision, dizziness, anxiety, impatience, and headaches [13]. Also, ketamine raises blood pressure and heart rate through sympathetic activation while maintaining respiratory activity, making a deadly overdose unlikely [13, 15]. Although the long-term safety profile of ketamine is unknown, it can cause bladder and urethral inflammation and irritation, and analogous changes in the biliary tract have recently been identified, resulting in acute or chronic cholestatic liver damage [4, 10, 15, 17]. Stopping the drug's use may help to reverse these adverse effects [17]. Madal et al.

[11] and Naughton et al. [10] found that the side effects were improved one hour after using ketamine in depressed patients.

7. Conclusion

Ketamine is a new and effective alternative drug for depression with a rapid beginning of action for the future. Slow intravenous ketamine treatment results in significant improvement in people with severe depression. However, there are still a number of gaps that remain, both in terms of clinical and research plans. In addition, the exact mechanism by which these antidepressant effects occur is still not fully resolved. We believe that future studies will shed light on new information on this subject.

Author details

Berzah Güneş[1*], Lora Koenhemsi[2] and Oytun Erbaş[3,4]

1 Department of Physiology, Demiroğlu Bilim University, Istanbul, Turkey

2 Faculty of Veterinary Medicine, Department of Internal Medicine, Istanbul University-Cerrahpaşa, Istanbul, Turkey

3 ERBAS Institute of Experimental Medicine, Gebze-Kocaeli, Türkiye

4 ERBAS Institute of Experimental Medicine, Illinois, USA

*Address all correspondence to: berzahgunes2020@gmail.com

References

[1] Goodwin RD, Dierker LC, Wu M, Galea S, Hoven CW, Weinberger AH. Trends in U.S depression prevalence from 2015 to 2020: The widening treatment gap. American Journal of Preventive Medicine. 2022;**63**(5):726-733. DOI: 10.1016/j.amepre.2022.05.014

[2] Yang T, Nie Z, Shu H, Kuang Y, Chen X, Cheng J, et al. The role of BDNF on neural plasticity in depression. Frontiers in Cellular Neuroscience. 2020;**14**:500839. DOI: 10.3389/fncel.2020.00082

[3] Kang MJ, Hawken E, Vazquez GH. The mechanisms behind rapid antidepressant effects of ketamine: A systematic review with a focus on molecular neuroplasticity. Frontiers. Psychiatry. 2022;**13**:1-17. DOI: 10.3389/fpsyt.2022.860882

[4] Zerzevatcı C, Kılıçoğlu RB, Genç AS, Erbaş O. Antidepressant effects of ketamine. Journal of Experimental and Basic Medical Sciences. 2022;**3**(3):227-231. DOI: 10.5606/jebms.2022.1032

[5] Zanos P, Gould TD. Mechanisms of ketamine action as an antidepressant. Molecular Psychiatry. 2018;**23**(4):801-811. DOI: 10.1038/mp.2017.255

[6] Meshkat S, Alnefeesi Y, Jawad MY, Di Vincenzo JD, Rodrigues NB, Ceban F, et al. Brain-derived neurotrophic factor (BDNF) as a biomarker of treatment response in patients with treatment resistant depression (TRD): A systematic review & meta-analysis. Psychiatry Research. 2022;**317**:114857. DOI: 10.1016/j.psychres.2022.114857

[7] Sforzini L, Worrell C, Kose M, Anderson AM, Aouizerate B, Arolt V, et al. A Delphi-method-based consensus guideline for definition of treatment-resistant depression for clinical trials. Molecular Psychiatry. 2022;**27**:1286-1299. DOI: 10.1038/s41380-021-01381-x

[8] Benson C, Szukis H, Sheehan JJ, Alphs L, Yuce H. An evaluation of the clinical and economic burden among older adult Medicare-covered beneficiaries with treatment-resistant depression. The American Journal of Geriatric Psychiatry. 2020;**28**(3):350-362. DOI: 10.1016/j.jagp.2019.10.012

[9] Garcia LS, Comim CM, Valvassori SS, Réus GZ, Andreazza AC, Stertz L, et al. Chronic administration of ketamine elicits antidepressant-like effects in rats without affecting hippocampal brain-derived neurotrophic factor protein levels. Basic and Clinical Pharmacology and Toxicology. 2008a;**103**:502-506. DOI: 10.1111/j.1742-7843.2008.00210.x

[10] Naughton M, Clarke G, O'Leary OF, Cryan JF, Dinan TG. A review of ketamine in affective disorders: Current evidence of clinical efficacy, limitations of use and pre-clinical evidence on proposed mechanisms of action. Journal of Affective Disorders. 2014;**156**:24-35. DOI: 10.1016/j.jad.2013.11.014

[11] Mandal S, Sinha VK, Goyal N. Efficacy of ketamine therapy in the treatment of depression. Indian Journal of Psychiatry. 2019;**61**(5):480-485. DOI: 10.4103/psychiatry.IndianJPsychiatry_484_18

[12] Marwaha S, Palmer E, Suppes T, Cons E, Young AH, Upthegrove R. Novel and emerging treatments for major depression. The Lancet. 2023;**401**(10371):141-153. DOI: 10.1016/S0140-6736(22)02080-3

[13] Serafini G, Howland RH, Rovedi F, Girardi P, Amore M. The role of ketamine in treatment-resistant depression: A systematic review. Current Neuropharmacology. 2014;**12**(5): 444-461. DOI: 10.2174/1570159X12666140619204251

[14] Woelfer M, Li M, Colic L, Liebe T, Di X, Biswal B, et al. Ketamine-induced changes in plasma brain-derived neurotrophic factor (BDNF) levels are associated with the resting-state functional connectivity of the prefrontal cortex. The World Journal of Biological Psychiatry. 2020;**21**(9):696-710. DOI: 10.1080/15622975.2019.1679391

[15] Peltoniemi MA, Hagelberg NM, Olkkola KT, Saari TI. Ketamine: A review of clinical pharmacokinetics and pharmacodynamics in anesthesia and pain therapy. Clinical Pharmacokinetics. 2016;**55**(9):1059-1077. DOI: 10.1007/s40262-016-0383-6

[16] Jelen LA, Stone JM. Ketamine for depression. International Review of Psychiatry. 2020;**33**(3):207-228. DOI: 10.1080/09540261.2020.1854194

[17] Li L, Vlisides PE. Ketamine: 50 years of modulating the mind. Frontiers in Human Neuroscience. 2016;**10**:1-15. DOI: 10.3389/fnhum.2016.00612

[18] Corssen G, Domino EF. Dissociative anesthesia: Further pharmacologic studies and first clinical experience with the phencyclidine derivative CI-581. Anesthesia and Analgesia. 1966;**45**(1):29-40. DOI: 10.1213/00000539-196601000-00007

[19] Kraus C, Wasserman D, Henter ID, Acevedo-Diaz E, Kadriu B, Zarate CA Jr. The influence of ketamine on drug discovery in depression. Drug Discovery Today. 2019;**24**(10):2033-2043. DOI: 10.1016/j.drudis.2019.07.007

[20] Khorramzadeh E, Lotfy AO. The use of ketamine in psychiatry. Psychosomatics. 1973;**14**(6):344-346. DOI: 10.1016/S0033-3182(73)71306-2

[21] Kolp E, Young MS, Friedman H, Krupitsky E, Jansen K, O'Connor LA. Ketamine-enhanced psychotherapy: Preliminary clinical observations on its effects in treating death anxiety. International Journal of Transpersonal Studies. 2007;**26**(1):1-17.b

[22] Berman RM, Cappiello A, Anand A, Oren DA, Heninger GR, Charney DS, et al. Antidepressant effects of ketamine in depressed patients. Biological Psychiatry. 2000;**47**(4):351-354. DOI: 10.1016/s0006-3223(99)00230-9

[23] Zarate CA, Singh JB, Carlson PJ, Brutsche NE, Ameli R, Luckenbaugh DA, et al. A randomized trial of an N-methyl-D-aspartate antagonist in treatment-resistant major depression. Archives of General Psychiatry. 2006;**63**(8):856-864. DOI: 10.1001/archpsyc.63.8.856

[24] Kayabaşı Y, Güneş B, Erbaş O. Serotonin receptors and depression. Journal of Experimental and Basic Medical Sciences. 2021;**2**(2):240-246. DOI: 10.5606/jebms.2021.75662

[25] Siuciak JA, Lewis DR, Wiegand SJ, Lindsay RM. Antidepressant-like effect of brain-derived neurotrophic factor (BDNF). Pharmacology Biochemistry and Behavior. 1997;**56**:131-137. DOI: 10.1016/S0091-3057(96)00169-4

[26] Duman RS, Monteggia LM. A neurotrophic model for stress-related mood disorder. Biological Psychiatry. 2006;**59**:1116-1127. DOI: 10.1016/j.biopsych.2006.02.013

[27] Duman RS, Li N. A neurotrophic hypothesis of depression: Role of synaptogenesis in the actions of NMDA

receptor antagonists. Philosophical Transactions of the Royal Society B. 2012;**367**:2475-2484. DOI: 10.1098/rstb.2011.0357

[28] Haile CN, Murrough JW, Iosifescu DV, Chang LC, Al Jurdi RK, Foulkes A, et al. Plasma brain derived neurotrophic factor (BDNF) and response to ketamine in treatment-resistant depression. The International Journal of Neuropsychopharmacology. 2014;**17**(2):331-336. DOI: 10.1017/S1461145713001119

[29] Karabaşoğlu C, Güneş B, Erbaş O. Suicide related genes. Journal of Experimental and Basic Medical Sciences. 2021;**2**(1):18-26. DOI: 10.5606/jebms.2021.75634

[30] Garcia LS, Comim CM, Valvassori SS, Réus GZ, Barbosa LM, Andreazza AC, et al. Acute administration of ketamine induces antidepressant-like effects in the forced swimming test and increases BDNF levels in the rat hippocampus. Progress in Neuro-Psychopharmacology & Biological Psychiatry. 2008b;**32**:140-144. DOI: 10.1016/j.pnpbp.2007.07.027

[31] Moghaddam B, Adams B, Verma A, Daly D. Activation of glutamatergic neurotransmission by ketamine: A novel step in the pathway from NMDA receptor blockade to dopaminergic and cognitive disruptions associated with the prefrontal cortex. The Journal of Neuroscience. 1997;**17**(8):2921-2927. DOI: 10.1523/JNEUROSCI

[32] Derkach VA, Oh MC, Guire ES, Soderling TR. Regulatory mechanisms of AMPA receptors in synaptic plasticity. Nature Reviews Neuroscience. 2007;**8**(2):101-113. DOI: 10.1038/nrn2055

[33] Rantamaki T, Hendolin P, Kankaanpaa A, Mijatovic J, Piepponen P, Domenici E, et al. Pharmacologically diverse antidepressants rapidly activate brain-derived neurotrophic factor receptor TrkB and induce phospholipase-Cgamma signaling pathways in mouse brain. Neuropsychopharmacology. 2007;**32**(10):2152-2162. DOI: 10.1038/sj.npp.1301345

[34] Yang C, Hu Y, Zhou Z, Zhang G, Yang J. Acute administration of ketamine in rats increases hippocampal BDNF and mTOR levels during forced swimming test. Upsala Journal of Medical Sciences. 2013;**118**(1):3-8. DOI: 10.3109/03009734.2012.724118

[35] Li N, Lee B, Liu RJ, Banasr M, Dwyer JM, Iwata M. mTOR-dependent synapse formation underlies the rapid antidepressant effects of NMDA antagonists. Science. 2010;**29**:959-964. DOI: 10.1126/science.1190287

[36] Carrier N, Kabbaj M. Sex differences in the antidepressant-like effects of ketamine. Neuropharmacology. 2013;**70**:27-34. DOI: 10.1016/j.neuropharm.2012.12.009

[37] Qu Y, Shan J, Wang S, Chang L, Pu Y, Wang X. Rapid-acting and long-lasting antidepressant-like action of (R)-ketamine in Nrf2 knock-out mice: A role of TrkB signaling. European Archives of Psychiatry and Clinical Neuroscience. 2021;**271**:439-446. DOI: 10.1007/s00406-020-01208-w

[38] Rafało-Ulińska A, Pałucha-Poniewiera A. The effectiveness of (R)-ketamine and its mechanism of action differ from those of (S)-ketamine in a chronic unpredictable mild stress model of depression in C57BL/6J mice. Behavioural Brain Research. 2022;**418**:113633. DOI: 10.1016/j.bbr.2021.113633

[39] Choi M, Lee SH, Park MH, Kim Y, Son H. Ketamine induces brain-derived

neurotrophic factor expression via phosphorylation of histone deacetylase 5 in rats. Biochemical and Biophysical Research Communications. 2017;**489**(4):420-425. DOI: 10.1016/j.bbrc.2017.05.157

[40] Tsankova NM, Berton O, Renthal W, Kumar A, Neve RL, Nestler EJ. Sustained hippocampal chromatin regulation in a mouse model of depression and antidepressant action. Nature Neuroscience. 2006;**9**:519-525. DOI: 10.1038/nn1659

[41] Renthal W, Maze I, Krishnan V, Covington HE, Xiao G, Kumar A, et al. Histone deacetylase 5 epigenetically controls Behavioral adaptations to chronic emotional stimuli. Neuron. 2007;**56**(3):517-529. DOI: 10.1016/j.neuron.2007.09.032

[42] Laje G, Lally N, Mathews D, Brutsche N, Chemerinski A, Akula N, et al. Brain-derived neurotrophic factor Val66Met polymorphism and antidepressant efficacy of ketamine in depressed patients. Biological Psychiatry. 2012;**72**(11):27-28. DOI: 10.1016/j.biopsych.2012.05.031

[43] Taha E, Gildish I, Gal-Ben-Ari S, Rosenblum K. The role of eEF2 pathway in learning and synaptic plasticity. Neurobiology of Learning and MemoryNeurobiology of Learning and Memory. 2013;**105**:100-106. DOI: 10.1016/j.nlm.2013.04.015

[44] Karege F, Schwald M, Cisse M. Postnatal developmental profile of brain-derived neurotrophic factor in rat brain and platelets. Neuroscience Letters. 2002;**328**(3):261-264. DOI: 10.1016/S03043940(02)00529-3

[45] Klein AB, Williamson R, Santini MA, Clemmensen C, Ettrup A, Rios M, et al. Blood BDNF concentrations reflect brain-tissue BDNF levels across species. The International Journal of Neuropsychopharmacology. 2011;**14**(03):347-353. DOI: 10.1017/S1461145710000738

[46] Lindholm JS, Autio H, Vesa L, Antila H, Lindemann L, Hoener MC. The antidepressant-like effects of glutamatergic drugs ketamine and AMPA receptor potentiator LY 451646 are preserved in bdnf+/− heterozygous null mice. Neuropharmacology. 2012;**62**(1):391-397. DOI: 10.1016/j.neuropharm.2011.08.015

Section 2

Brain-Derived Neurotrophic Factor for Treatment of Spinal Cord Injury

Chapter 4

Neuromodulatory Effect of BDNF in Spinal Cord Injury

Mehmet Burak Yalçın

Abstract

The neuromodulatory effect of brain-derived neurotrophic factor (BDNF) in spinal cord injury (SCI) is a topic of significant interest. BDNF, a neurotrophic factor, plays a crucial role in promoting neuronal survival, axonal growth, and synaptic plasticity in the central nervous system. In SCI, BDNF has been shown to enhance the survival of injured neurons and stimulate axonal growth through the activation of downstream signaling pathways. Additionally, BDNF exhibits potent anti-inflammatory effects, reducing neuroinflammation and secondary damage. The timing and duration of BDNF administration are critical, with early intervention showing better outcomes. However, the optimal dosage and frequency of BDNF administration remain to be determined. Further research is needed to fully understand the potential of BDNF as a therapeutic agent for enhancing functional recovery and promoting neuroplasticity in individuals with SCI.

Keywords: spinal cord injury, BDNF, neuromodulation, orthopedics, remodeling

1. Introduction

Spinal cord injury (SCI) is a catastrophic condition that impacts millions of people globally, causing permanent loss of motor and sensory function below the injury site, leading to a diminished quality of life [1]. Despite advancements in medical care and rehabilitation, there are no successful treatments available to promote substantial functional recovery following SCI.

Neurotrophic factors, such as brain-derived neurotrophic factor (BDNF), have emerged as promising therapeutic agents for promoting functional recovery after SCI [2]. BDNF is a member of the neurotrophin family of growth factors and plays a crucial role in promoting neuronal survival, axonal growth, and synaptic plasticity in the central nervous system (CNS). In addition to its neurotrophic effects, BDNF also has potent anti-inflammatory and neuroprotective effects, making it an attractive candidate for promoting functional recovery after SCI [3] (**Figure 1**).

Several preclinical and clinical studies have demonstrated the potential of BDNF for promoting functional recovery after SCI [3–5]. These studies have shown that BDNF administration can promote axonal regeneration, improve synaptic plasticity, and reduce inflammation in animal models of SCI and in humans [4]. Despite these promising results, there are still many questions that remain unanswered regarding

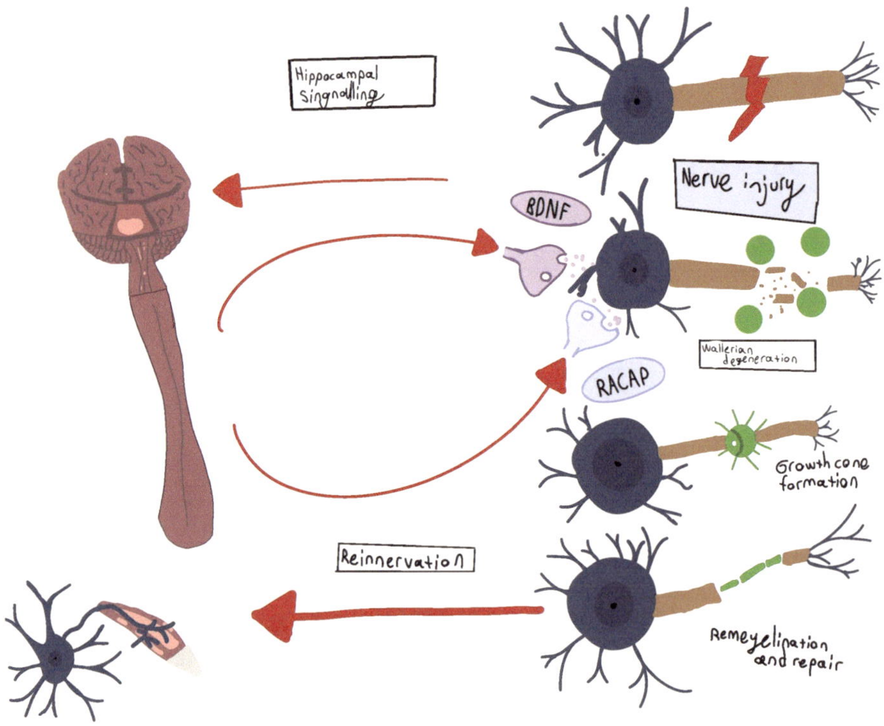

Figure 1.
Relation between BDNF and axonal regeneration (Thank you to Derin Mavi Bora for the illustration).

the optimal timing, dose, and route of administration of BDNF for SCI, as well as the long-term safety and efficacy of BDNF in humans.

In this review, we will explore the neuromodulatory effects of BDNF in SCI, including its effects on axonal regeneration, synaptic plasticity, and inflammation. We will also discuss the challenges and limitations of using BDNF for SCI, as well as future directions for research in this area. By examining the current state of knowledge on the neuromodulatory effects of BDNF in SCI, we hope to provide a comprehensive overview of this promising therapeutic approach and its potential for promoting functional recovery after SCI.

2. Axonal regeneration

Axonal regeneration is a critical process for promoting functional recovery after SCI. However, the regenerative capacity of the central nervous system (CNS) is limited, and axonal regeneration after SCI is typically minimal. BDNF has been shown to promote axonal regeneration in animal models of SCI through several mechanisms [6].

BDNF initially enhances the survival of damaged neurons and facilitates the growth of axons. The protein induces this effect by activating signaling pathways downstream, which control the expression of genes linked to axonal growth and survival. These signaling pathways involve the mitogen-activated protein kinase (MAPK) and phosphoinositide 3-kinase (PI3K) pathways, which regulate the production of growth-associated proteins and encourage the growth and survival of axons [7].

Second, BDNF promotes the migration of neural stem cells (NSCs) to the site of injury, where they can differentiate into neurons and support neuronal survival and growth [7]. The protein achieves this effect by activating downstream signaling pathways that regulate the migration and differentiation of NSCs, including the MAPK pathway and the PI3K pathway [8].

Third, BDNF promotes the formation of new synapses and the reorganization of neural circuits. The protein achieves this effect by activating downstream signaling pathways that regulate the expression of synaptic proteins and promote the formation of new synapses. These pathways include the protein kinase B (AKT) pathway and the extracellular signal-regulated kinase (ERK) pathway, which regulate the expression of synaptic proteins and promote the formation of new synapses [9].

Axonal regeneration refers to the process by which damaged or severed axons in the nervous system grow and reestablish connections with their target tissues. In the case of spinal cord injury (SCI), axonal regeneration is a critical step in promoting functional recovery as it can help restore communication between the brain and the rest of the body [6].

Axonal regeneration is a complex process that involves several steps. First, the damaged axon must form a growth cone, which is a specialized structure at the tip of the axon that guides its growth toward the target tissue. The growth cone is sensitive to guidance cues in the extracellular environment, such as extracellular matrix molecules and chemotropic factors, which help direct the axon toward its target [8].

Once the growth cone reaches its target tissue, it must form a new synapse, which is the site of communication between the axon and the target cell. Synapse formation requires the release of neurotransmitters from the axon and the activation of receptors on the target cell, leading to the transmission of signals between the two cells [8].

In the context of SCI, axonal regeneration is hindered by several factors, including the presence of inhibitory factors in the extracellular environment, the loss of trophic support from target tissues, and the formation of a glial scar, which is a dense network of astrocytes that form at the site of injury and inhibit axonal regeneration.

BDNF has been shown to promote axonal regeneration after SCI by overcoming some of these inhibitory factors and promoting the growth and guidance of damaged axons. BDNF can promote axonal regeneration by increasing the expression of growth cone proteins and chemotropic factors, reducing the expression of inhibitory factors, and enhancing the survival and growth of damaged axons [9]. By promoting axonal regeneration, BDNF can help restore communication between the brain and the rest of the body, leading to improved functional recovery after SCI [7].

3. Neuronal excitability

Neuronal excitability is a critical process for promoting functional recovery after SCI. After SCI, there is a significant decrease in neuronal excitability, which contributes to the loss of neurological function. BDNF has been shown to promote neuronal excitability in animal models of SCI through several mechanisms [10].

First, BDNF promotes the expression of ion channels and receptors that regulate neuronal excitability [11]. The protein achieves this effect by activating downstream signaling pathways that regulate the transcription and translation of ion channels and receptors, including the AKT pathway and the ERK pathway [12].

Second, BDNF modulates synaptic transmission by regulating the release of neurotransmitters and the expression of neurotransmitter receptors. The protein achieves this effect by activating downstream signaling pathways that regulate the expression and trafficking of neurotransmitter receptors [13].

Third, BDNF promotes the generation and propagation of action potentials by regulating the expression and activity of voltage-gated ion channels [13].

4. Synaptic plasticity

Synaptic plasticity is a critical process for promoting functional recovery after SCI. It is the ability of synapses to change their strength and efficacy in response to activity, which is essential for learning, memory, and adaptive behavior. After SCI, there is a significant decrease in synaptic plasticity, which contributes to the loss of neurological function. BDNF has been shown to promote synaptic plasticity in animal models of SCI through several mechanisms [14].

First, BDNF promotes the expression and trafficking of AMPA receptors, which are critical for synaptic plasticity. The protein achieves this effect by activating downstream signaling pathways that regulate the expression and trafficking of AMPA receptors, including the AKT pathway and the ERK pathway [15].

Second, BDNF promotes the formation and stabilization of dendritic spines, which are the primary sites of excitatory synaptic transmission. The protein achieves this effect by activating downstream signaling pathways that regulate the expression of cytoskeletal proteins and promote the formation and stabilization of dendritic spines [15].

Third, BDNF promotes the release of neurotransmitters and the activation of downstream signaling pathways that regulate synaptic plasticity. The protein achieves this effect by activating downstream signaling pathways that regulate the release of neurotransmitters, including the MAPK pathway and the PI3K pathway [16].

Synaptic plasticity refers to the ability of synapses, the connections between neurons in the nervous system, to change and adapt in response to experience or injury. Synaptic plasticity is a fundamental process underlying learning, memory, and recovery from injury in the nervous system.

In the context of spinal cord injury (SCI), synaptic plasticity can play a critical role in promoting functional recovery by allowing for the formation of new connections and the strengthening of existing ones. After SCI, the loss of connections between neurons can lead to a reduction in synaptic activity, which can impair motor and sensory function [14].

BDNF is a potent modulator of synaptic plasticity in the nervous system and has been shown to promote the formation of new synapses, increase the strength of existing ones, and enhance synaptic transmission after injury. BDNF can promote synaptic plasticity by increasing the release of neurotransmitters, enhancing the expression and localization of synaptic proteins, and regulating synaptic pruning and remodeling [17].

BDNF's ability to promote synaptic plasticity after SCI makes it a promising therapeutic agent for promoting functional recovery. By enhancing synaptic plasticity,

BDNF can help restore connectivity between neurons, leading to improved motor and sensory function after SCI [17].

In addition to its effects on synaptic plasticity, BDNF can also promote neuronal survival and protect against apoptosis, which can further contribute to functional recovery after SCI. By promoting neuronal survival and protecting against cell death, BDNF can help maintain the integrity of neural circuits and prevent the loss of connections between neurons.

Overall, synaptic plasticity is a critical process underlying functional recovery after SCI, and BDNF's ability to enhance synaptic plasticity and promote neuronal survival makes it a promising therapeutic agent for promoting functional recovery in this population.

5. Neuroinflammation

Neuroinflammation is a crucial process that occurs following SCI and is responsible for the decline in neurological function. It involves the activation of microglia and astrocytes, the release of proinflammatory cytokines, and the recruitment of immune cells to the site of injury. However, BDNF has been demonstrated to possess anti-inflammatory effects in animal models of SCI through various mechanisms [18].

First, BDNF reduces the activation of microglia and astrocytes, which are the primary cells responsible for neuroinflammation. The protein achieves this effect by activating downstream signaling pathways that regulate the expression of anti-inflammatory cytokines and reduce the activation of microglia and astrocytes.

Second, BDNF reduces the release of proinflammatory cytokines, which contribute to the progression of neuroinflammation. The protein achieves this effect by activating downstream signaling pathways that regulate the expression of anti-inflammatory cytokines and reduce the release of proinflammatory cytokines [14].

Third, BDNF promotes the recruitment of immune cells that have anti-inflammatory effects, such as regulatory T cells and M2 macrophages. The protein achieves this effect by activating downstream signaling pathways that regulate the recruitment and activation of immune cells.

Neuroinflammation is a multifaceted process that involves the activation of immune cells in the CNS as a reaction to injury or illness. In the context of spinal cord injury (SCI), neuroinflammation is a significant characteristic and is correlated with secondary damage to spinal cord tissue, including neuronal demise, demyelination, and axonal harm [19].

After SCI, activated microglia and infiltrating immune cells release proinflammatory cytokines and chemokines, leading to the recruitment of additional immune cells to the site of injury. The resulting immune response can exacerbate tissue damage and contribute to the development of a glial scar, which inhibits axonal regeneration and functional recovery [19].

BDNF has been shown to have potent anti-inflammatory effects in the CNS, which can help mitigate the detrimental effects of neuroinflammation after SCI. BDNF can reduce the production of proinflammatory cytokines and chemokines, decrease microglial activation, and promote the polarization of microglia toward an anti-inflammatory phenotype. By reducing neuroinflammation, BDNF can help promote tissue repair and functional recovery after SCI.

Furthermore, BDNF can also promote the survival of neurons and glial cells in the CNS, which can further reduce neuroinflammation by preventing the release of damage-associated molecular patterns (DAMPs) that activate the immune response. By promoting cell survival and reducing the release of DAMPs, BDNF can help prevent the perpetuation of neuroinflammation after SCI [20].

Overall, neuroinflammation is a complex process that can contribute to secondary damage and hinder functional recovery after SCI. BDNF's anti-inflammatory effects make it a promising therapeutic agent for mitigating the detrimental effects of neuroinflammation and promoting tissue repair and functional recovery after SCI [21].

6. Clinical implications

The neuromodulatory effects of BDNF in SCI have significant clinical implications for the treatment of this devastating neurological condition. Several studies have shown that the administration of exogenous BDNF can promote axonal regeneration, neuronal excitability, synaptic plasticity, and anti-inflammatory effects in animal models of SCI [22–24]. However, translating these findings into clinical practice presents several challenges.

First, BDNF is a large protein that does not readily cross the blood-brain barrier, which limits its effectiveness as a therapeutic agent. Several strategies have been developed to overcome this limitation, including the use of viral vectors to deliver BDNF directly to the site of injury and the use of small molecule agonists of BDNF receptors [16].

Second, the timing and duration of BDNF administration are essential factors for promoting axonal regeneration and functional recovery following SCI. Multiple studies have demonstrated that administering BDNF early after SCI can enhance axonal regeneration and functional recovery, while delayed administration may have a reduced effect. Nevertheless, the ideal duration and frequency of BDNF administration are not yet well established [15, 25, 26].

Third, the potential side effects of BDNF administration are not well understood. Several studies have shown that BDNF administration can promote tumor growth and metastasis in animal models, which raises concerns about its safety for clinical use. However, these findings have not been replicated in human studies, and further research is needed to evaluate the safety profile of BDNF in humans [24, 27, 28].

Despite these challenges, several clinical trials have been conducted to evaluate the efficacy of BDNF in promoting functional recovery after SCI. One study conducted in China evaluated the safety and efficacy of intrathecal administration of recombinant human BDNF in patients with complete SCI. The results showed that BDNF administration was well tolerated and resulted in significant improvements in neurological function, including motor and sensory function, bladder function, and spasticity.

Another study conducted in the United States evaluated the safety and efficacy of intrathecal administration of a viral vector encoding BDNF in patients with chronic SCI. The results showed that BDNF administration was well tolerated and resulted in significant improvements in neurological function, including motor and sensory function, bladder function, and quality of life.

7. Conclusion

SCI is a devastating neurological condition that results in significant loss of neurological function. The neuromodulatory effects of BDNF in SCI have significant potential for promoting axonal regeneration, synaptic plasticity, and anti-inflammatory effects, which can lead to functional recovery after injury. Several preclinical and clinical studies have shown that BDNF administration can promote functional recovery in animal models and humans, but further research is needed to optimize the timing, duration, and frequency of BDNF administration and evaluate its safety profile in humans. Despite these challenges, the potential benefits of BDNF administration for promoting functional recovery after SCI are significant and warrant further investigation.

8. Limitations

In addition to the challenges discussed above, there are several other limitations and potential areas for improvement in the use of BDNF for SCI. One limitation is the heterogeneity of SCI patients, which can lead to variability in treatment response and may require personalized treatment strategies. For example, patients with different injury levels, severity, and comorbidities may require different doses or timing of BDNF administration, and response to treatment may depend on individual factors such as age, sex, and genetics. Therefore, future studies should aim to identify patient-specific factors that can predict treatment response and guide personalized treatment strategies.

Another potential area for improvement is the development of biomarkers for monitoring treatment response and predicting outcomes. Currently, there are no reliable biomarkers for monitoring the efficacy of BDNF treatment or predicting long-term outcomes after SCI. Developing biomarkers that can measure the extent of axonal regeneration, synaptic plasticity, or inflammation in response to BDNF treatment could provide valuable information for optimizing treatment strategies and predicting outcomes.

Finally, there is a need for improved animal models of SCI that can better mimic the complex pathophysiology of human SCI. Currently, most preclinical studies use rodent models of SCI, which have several limitations in terms of size, anatomy, and physiology compared to humans. Developing larger animal models, such as nonhuman primates or canines, that more closely resemble human SCI could provide more accurate and translational data for guiding clinical trials.

Despite these challenges and limitations, the neuromodulatory effects of BDNF in SCI hold significant promise for promoting functional recovery after injury. Further research is needed to optimize treatment strategies, evaluate safety and efficacy, and address remaining questions and limitations. Ultimately, the use of BDNF and other neurotrophic factors for SCI represents an exciting area of research with the potential to improve the lives of millions of individuals living with SCI.

Conflict of interest

The authors declare no conflict of interest.

Thanks

Special thanks to Assoc. Prof. MD, PhD Ejder Saylav Bora (Izmir Ataturk Research and Training Hospital) for his scientific supports.

Author details

Mehmet Burak Yalçın
Department of Orthopedics and Traumatology, Bahcelievler Memorial Hospital, Istanbul, Türkiye

*Address all correspondence to: mehmetburakyalcin@gmail.com

References

[1] Eli I, Lerner DP, Ghogawala Z. Acute traumatic spinal cord injury. Neurologic Clinics. 2021;**39**(2):471-488. DOI: 10.1016/j.ncl.2021.02.004

[2] Głowacka A, Ji B, Szczepankiewicz AA, Skup M, Gajewska-Woźniak O. BDNF spinal overexpression after spinal cord injury partially protects soleus neuromuscular junction from disintegration, increasing VAChT and AChE transcripts in soleus but not Tibialis anterior Motoneurons. Biomedicine. 2022;**10**(11):2851. DOI: 10.3390/biomedicines10112851

[3] Colucci-D'Amato L, Speranza L, Volpicelli F. Neurotrophic factor BDNF, physiological functions and therapeutic potential in depression, neurodegeneration and brain cancer. International Journal of Molecular Sciences. 2020;**21**(20):7777. DOI: 10.3390/ijms21207777

[4] Otsuka S, Sakakima H, Sumizono M, Takada S, Terashi T, Yoshida Y. The neuroprotective effects of preconditioning exercise on brain damage and neurotrophic factors after focal brain ischemia in rats. Behavioural Brain Research. 2016;**303**:9-18. DOI: 10.1016/j.bbr.2016.01.049

[5] Sayyah M, Seydyousefi M, Moghanlou AE, Metz GAS, Shamsaei N, Faghfoori MH, et al. Activation of BDNF- and VEGF-mediated neuroprotection by treadmill exercise training in experimental stroke. Metabolic Brain Disease. 2022;**37**(6):1843-1853. DOI: 10.1007/s11011-022-01003-7

[6] Han Q, Xie Y, Ordaz JD, et al. Restoring cellular energetics promotes axonal regeneration and functional recovery after spinal cord injury. Cell Metabolism. 2020;**31**(3):623-641.e8. DOI: 10.1016/j.cmet.2020.02.002

[7] Godinho MJ, Teh L, Pollett MA, Goodman D, Hodgetts SI, Sweetman I, et al. Immunohistochemical, ultrastructural and functional analysis of axonal regeneration through peripheral nerve grafts containing Schwann cells expressing BDNF, CNTF or NT3. PLoS One. 2013;**8**(8):e69987. DOI: 10.1371/journal.pone.0069987

[8] Akram R, Anwar H, Javed MS, Rasul A, Imran A, Malik SA, et al. Axonal regeneration: Underlying molecular mechanisms and potential therapeutic targets. Biomedicine. 2022;**10**(12):3186. DOI: 10.3390/biomedicines10123186

[9] Chen X, You S, Xia Q, Mo X. 7,8-Dihydroxiflavone protects retinal ganglion cells and promotes axonal regeneration through TrkB signaling pathway followed by AKT and ERK activation. Neuroscience Research. 2023;**23**:00050. Advance online publication. DOI: 10.1016/j.neures.2023.02.008

[10] Abidin İ, Aydin-Abidin S, Mittmann T. Neuronal excitability and spontaneous synaptic transmission in the entorhinal cortex of BDNF heterozygous mice. Neuroscience Letters. 2019;**690**:69-75. DOI: 10.1016/j.neulet.2018.10.019

[11] Boakye PA, Tang SJ, Smith PA. Mediators of neuropathic pain; focus on spinal microglia, CSF-1, BDNF, CCL21, TNF-α, Wnt Ligands, and Interleukin 1β. Frontier in Pain Research. 2021;**2**:698157. DOI: 10.3389/fpain.2021.698157

[12] Zhang C, Yuan XR, Li HY, Zhao ZJ, Liao YW, Wang XY, et al. Anti-cancer effect of metabotropic glutamate

receptor 1 inhibition in human glioma U87 cells: Involvement of PI3K/Akt/mTOR pathway. Cellular Physiology and Biochemistry : International Journal of Experimental Cellular Physiology, Biochemistry, and Pharmacology. 2015;**35**(2):419-432. DOI: 10.1159/000369707

[13] Sun W, Lu Z, Chen X, Yang Mei Y, Li X, An L. Aluminum oxide nanoparticles impair working memory and neuronal activity through the GSK3β/BDNF Signaling pathway of prefrontal cortex in rats. ACS Chemical Neuroscience. 2022;**13**(23):3352-3361. DOI: 10.1021/acschemneuro.2c00383

[14] Leal G, Comprido D, Duarte CB. BDNF-induced local protein synthesis and synaptic plasticity. Neuropharmacology. 2014;**76 Pt C**:639-656. DOI: 10.1016/j.neuropharm.2013.04.005

[15] Rosa PB, Bettio LEB, Neis VB, Moretti M, Kaufmann FN, Tavares MK, et al. Antidepressant-like effect of guanosine involves activation of AMPA receptor and BDNF/TrkB signaling. Purinergic Signalling. 2021;**17**(2):285-301. DOI: 10.1007/s11302-021-09779-6

[16] Lu B, Chow A. Neurotrophins and hippocampal synaptic transmission and plasticity. Journal of Neuroscience Research. 1999;**58**(1):76-87

[17] Lu B, Nagappan G, Lu Y. BDNF and synaptic plasticity, cognitive function, and dysfunction. Handbook of Experimental Pharmacology. 2014;**220**:223-250. DOI: 10.1007/978-3-642-45106-5_9

[18] Ding H, Chen J, Su M, et al. BDNF promotes activation of astrocytes and microglia contributing to neuroinflammation and mechanical allodynia in cyclophosphamide-induced cystitis. Journal of Neuroinflammation. 2020;**17**:19. DOI: 10.1186/s12974-020-1704-0

[19] Leng F, Edison P. Neuroinflammation and microglial activation in Alzheimer disease: Where do we go from here? Nature Reviews. Neurology. 2021;**17**(3):157-172. DOI: 10.1038/s41582-020-00435-y

[20] Gong T, Liu L, Jiang W, Zhou R. DAMP-sensing receptors in sterile inflammation and inflammatory diseases. Nature Reviews. Immunology. 2020;**20**(2):95-112. DOI: 10.1038/s41577-019-0215-7

[21] Castrén E, Monteggia LM. Brain-derived neurotrophic factor Signaling in depression and antidepressant action. Biological Psychiatry. 2021;**90**(2):128-136. DOI: 10.1016/j.biopsych.2021.05.008

[22] Khayatan D, Razavi SM, Arab ZN, Niknejad AH, Nouri K, Momtaz S, et al. Protective effects of curcumin against traumatic brain injury. Biomedicine & Pharmacotherapy = Biomedecine & Pharmacotherapie. 2022;**154**:113621. DOI: 10.1016/j.biopha.2022.113621

[23] Bathina S, Das UN. Brain-derived neurotrophic factor and its clinical implications. Archives of Medical Science: AMS. 2015;**11**(6):1164-1178. DOI: 10.5114/aoms.2015.56342

[24] Huie JR, Garraway SM, Baumbauer KM, Hoy KC Jr, Beas BS, Montgomery KS, et al. Brain-derived neurotrophic factor promotes adaptive plasticity within the spinal cord and mediates the beneficial effects of controllable stimulation. Neuroscience. 2012;**200**:74-90. DOI: 10.1016/j.neuroscience.2011.10.028

[25] Zheng J, Sun J, Lu X, Zhao P, Li K, Li L. BDNF promotes the axonal

regrowth after sciatic nerve crush through intrinsic neuronal capability upregulation and distal portion protection. Neuroscience Letters. 2016;**621**:1-8. DOI: 10.1016/j.neulet.2016.04.006

[26] Blaszkiewicz M, Gunsch G, Willows JW, Gardner ML, Sepeda JA, Sas AR, et al. Adipose tissue myeloid-lineage Neuroimmune cells express genes important for neural plasticity and regulate adipose innervation. Frontiers in Endocrinology. 2022;**13**:864925. DOI: 10.3389/fendo.2022.864925

[27] Weishaupt N, Blesch A, Fouad K. BDNF: The career of a multifaceted neurotrophin in spinal cord injury. Experimental Neurology. 2012;**238**(2):254-264. DOI: 10.1016/j.expneurol.2012.09.001

[28] Tekgul H, Simsek E, Erdoğan MA, Yiğittürk G, Erbaş O, Taşkıran D. The potential effects of anticonvulsant drugs on neuropeptides and neurotrophins in pentylenetetrazol kindled seizures in the rat. The International Journal of Neuroscience. 2020;**130**(2):193-203. DOI: 10.1080/00207454.2019.1667791

Section 3

Sepsis and Brain-Derived Neurotrophic Factor

Chapter 5

Sepsis and Brain-Derived Neurotrophic Factor (BDNF): Exploring the Complex Connection

Ejder Saylav Bora

Abstract

In recent studies, brain-derived neurotrophic factor (BDNF) become a very important position. Because it is now known that it is not just a hormone that is released from the hippocampus and which supports the differentiation and growth of newly formed nerve cells and synapses while maintaining the vitality of existing neurons. Today BDNF was used as an indicator of severe sepsis and also in the follow-up of the disease. Moreover, BDNF is a potential anti-inflammatory agent which can be given like a medicament. In some studies, antiinflammatory effect was proven "in acute lung injury, in myocardial injury, in hepatorenal injury" triggered by sepsis. In this chapter, we will try to explain the BDNF effect in sepsis according to recent literature and update our knowledge.

Keywords: sepsis, BDNF, Antiinflammation, multiorgan failure, oxidative stress, biomarker

1. Introduction

Sepsis is a life-threatening condition caused by a dysregulated response to infection that can lead to organ dysfunction and failure [1]. Despite advances in medical care, sepsis remains a significant global health concern. Researchers have been investigating various aspects of sepsis pathophysiology to improve understanding and identify potential therapeutic targets [2]. One such area of exploration is the role of brain-derived neurotrophic factor (BDNF), a key protein involved in neuronal survival, growth, and plasticity [3]. Recent studies suggest that BDNF may play a crucial role in sepsis-associated brain dysfunction, contributing to long-term cognitive impairment observed in septic patients [4, 5]. This review aims to delve into the complex connection between sepsis and BDNF, discussing the underlying mechanisms, clinical implications, and potential therapeutic interventions.

IntechOpen

2. Sepsis: A brief overview

2.1 Definition and prevalence of sepsis

Sepsis is a life-threatening condition characterized by a dysregulated immune response to an infection. It occurs when the body's response to infection becomes overwhelming, leading to widespread inflammation and organ dysfunction [1]. Sepsis can progress rapidly and result in septic shock, a severe form of the condition associated with dangerously low blood pressure and inadequate blood flow to vital organs. The prevalence of sepsis is significant worldwide, with millions of cases reported each year [1, 2]. It affects individuals of all ages, but the elderly, young children, and those with weakened immune systems are particularly vulnerable.

2.2 Pathophysiology and immune response in sepsis

The pathophysiology of sepsis involves a complex interplay between the immune system, inflammatory mediators, and invading pathogens. When an infection occurs, the immune system initiates a response to control and eliminate the pathogens [4]. However, in sepsis, the immune response becomes dysregulated, leading to an excessive release of pro-inflammatory cytokines and the activation of immune cells. This immune activation triggers a cascade of events that can result in damage to organs and tissues throughout the body [2, 5].

2.3 Clinical manifestations and complications of sepsis

The clinical manifestations of sepsis can vary widely, making early diagnosis challenging [5]. Common signs and symptoms include fever, increased heart rate, rapid breathing, and altered mental status [4]. As sepsis progresses, patients may experience organ dysfunction, such as respiratory failure, acute kidney injury, or cardiovascular collapse. If septic shock develops, additional complications can arise, including multiple organ failure and disseminated intravascular coagulation (DIC), a condition characterized by abnormal blood clotting [6].

The complications associated with sepsis can have long-lasting effects on patients' health and quality of life. Survivors of sepsis may experience physical, cognitive, and psychological impairments. Cognitive dysfunction, often referred to as sepsis-associated encephalopathy (SAE), is a common neurological complication characterized by confusion, memory loss, and difficulty concentrating. Moreover, sepsis survivors may be at an increased risk of developing post-sepsis syndrome, a condition characterized by persistent fatigue, muscle weakness, and mood disturbances [4, 7].

In conclusion, sepsis is a severe and life-threatening condition characterized by a dysregulated immune response to infection. Its pathophysiology involves a complex interplay of immune mediators, leading to widespread inflammation and organ dysfunction. Prompt recognition and early intervention are critical to improving patient outcomes. The clinical manifestations of sepsis can be diverse, and its complications can have long-term effects on survivors. Further research and advancements in sepsis management are necessary to reduce its burden and improve patient care.

3. Introduction to brain-derived neurotrophic factor (BDNF)

3.1 BDNF: Structure, synthesis, and function

BDNF is a protein that belongs to the neurotrophin family. It is widely expressed in the central nervous system, including the brain and spinal cord. BDNF is synthesized as a precursor molecule called proBDNF, which is then cleaved to form mature BDNF. The mature form of BDNF is secreted and acts upon specific receptors to exert its biological effects [8].

BDNF plays a crucial role in promoting the survival, growth, and maintenance of neurons. It supports neuronal function and plasticity by influencing various cellular processes, including synaptic transmission, dendritic growth, and neuronal connectivity. BDNF also has neuroprotective properties and can modulate neuronal responses to injury and stress [9].

3.2 DNF and neuronal development

During early development, BDNF is involved in guiding the formation and connectivity of neurons. It promotes neuronal survival and influences the growth and branching of dendrites and axons. BDNF is particularly important in the development of the central nervous system, including the formation of neural circuits and the establishment of synaptic connections [9].

Studies have shown that BDNF plays a critical role in neurogenesis, the process of generating new neurons. It regulates the proliferation, differentiation, and survival of neural stem cells and progenitor cells. By promoting the production and integration of new neurons, BDNF contributes to the plasticity and adaptability of the developing brain [10].

3.3 Role of BDNF in synaptic plasticity and cognition

Synaptic plasticity refers to the ability of synapses to modify their strength and connectivity in response to activity and experience. BDNF is a key player in synaptic plasticity, particularly in long-term potentiation and long-term depression, which are fundamental processes underlying learning and memory [10].

BDNF promotes the formation and stabilization of synapses, enhances synaptic transmission, and modulates the structural and functional properties of synapses. It acts by binding to its specific receptor, tropomyosin receptor kinase B (TrkB), and activating intracellular signaling pathways that lead to changes in gene expression and neuronal function [11, 12].

In the context of cognition, BDNF is crucial for various forms of learning and memory. Studies have demonstrated that BDNF levels increase during learning tasks, and disruptions in BDNF signaling can impair cognitive function. BDNF influences the synaptic changes necessary for memory formation and retrieval, and it is involved in the maintenance of cognitive processes such as attention, executive function, and synaptic plasticity [9–12].

In summary, BDNF is a vital protein involved in neuronal development, synaptic plasticity, and cognitive processes. It supports the survival and growth of neurons, guides neuronal connectivity during development, and plays a key role in synaptic plasticity and memory formation. The intricate functions of BDNF make it a compelling candidate for investigating its involvement in sepsis-associated brain dysfunction (**Figure 1**).

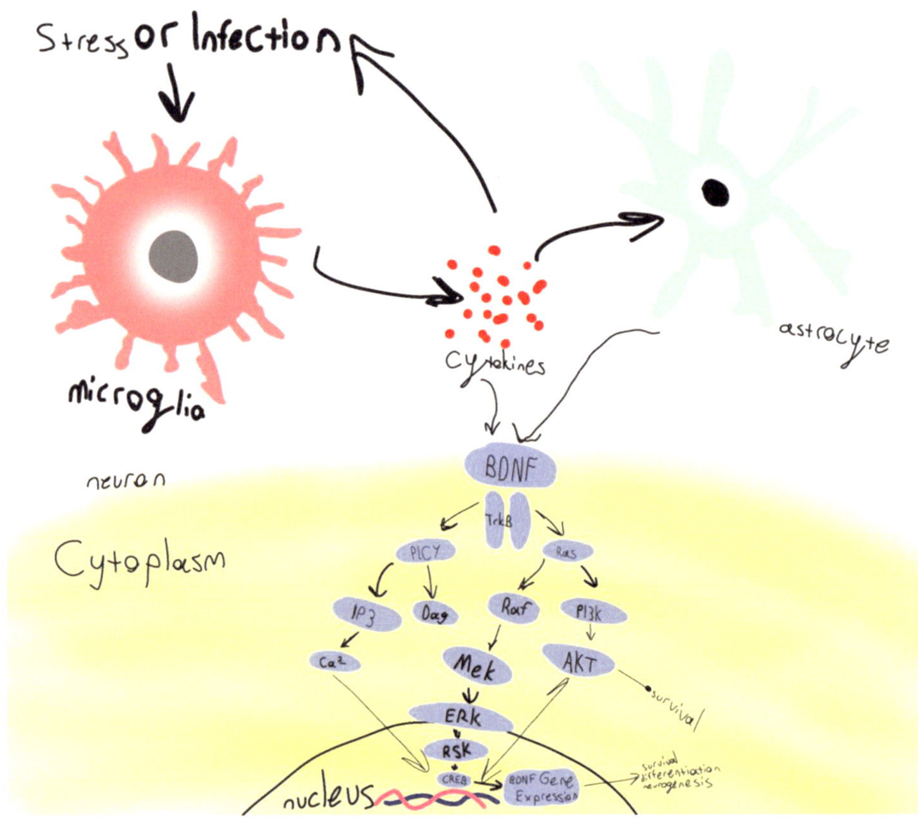

Figure 1.
The role of BDNF in the Neuroimmune Axis regulation (thank you to Derin Mavi bora for her support in this figure).

4. Sepsis-associated brain dysfunction

4.1 Understanding sepsis-associated encephalopathy (SAE)

SAE refers to the neurological dysfunction and cognitive impairment observed in patients with sepsis [13]. It is a common complication of sepsis, affecting a significant proportion of patients. SAE is characterized by a range of cognitive deficits, including confusion, delirium, memory impairment, attention deficits, and alterations in consciousness [13, 14]. It can have a significant impact on patient outcomes and contribute to long-term cognitive impairment.

The exact mechanisms underlying SAE are not fully understood, but several factors likely contribute to its development. These include the direct effects of the infectious agents or their byproducts, the systemic inflammatory response, and the impact of altered cerebral blood flow and oxygenation [14, 15]. The multifactorial nature of SAE makes it a complex condition to study and manage effectively.

4.2 Mechanisms of brain injury in sepsis

Sepsis can lead to brain injury through various mechanisms. The systemic inflammatory response in sepsis triggers the release of pro-inflammatory cytokines and other inflammatory mediators, which can cross the blood-brain barrier and induce neuroinflammation [15]. Neuroinflammation contributes to the disruption of normal brain function and can lead to neuronal damage and death.

Another mechanism of brain injury in sepsis is the dysregulation of cerebral blood flow [16, 17]. Sepsis can result in abnormal vascular function, including microvascular dysfunction and impaired autoregulation. These alterations in blood flow can lead to hypoxia and ischemia in the brain, causing neuronal injury.

Additionally, sepsis-induced oxidative stress plays a crucial role in brain injury. Increased production of reactive oxygen species (ROS) overwhelms the antioxidant defense mechanisms, leading to oxidative damage to neuronal cells [18]. Oxidative stress can impair cellular structures, disrupt neurotransmitter balance, and promote neuroinflammation, ultimately contributing to cognitive dysfunction in septic patients.

4.3 Neuroinflammation and oxidative stress in septic brains

Neuroinflammation and oxidative stress are closely intertwined processes that play significant roles in sepsis-associated brain dysfunction. The release of pro-inflammatory cytokines, such as interleukin-1β (IL-1β) and tumor necrosis factor-α (TNF-α), activates resident immune cells in the brain, such as microglia, leading to inflammatory response [16, 17]. Activated microglia produce additional pro-inflammatory mediators, perpetuating neuroinflammation.

Neuroinflammation can disrupt the delicate balance of neurotransmitters in the brain, impair synaptic transmission, and contribute to neuronal dysfunction. It also activates signaling pathways that induce the expression of enzymes that generate reactive oxygen species, exacerbating oxidative stress [18]. Oxidative stress, in turn, leads to lipid peroxidation, protein oxidation, and DNA damage in neuronal cells, further compromising their function and viability [19].

The combination of neuroinflammation and oxidative stress creates a detrimental cycle in septic brains, leading to progressive brain injury and cognitive impairment [20]. The sustained activation of inflammatory responses and the accumulation of oxidative damage contribute to the long-term consequences of sepsis-associated brain dysfunction [21].

In summary, sepsis-associated brain dysfunction involves complex mechanisms of brain injury, including neuroinflammation and oxidative stress. The systemic inflammatory response, altered cerebral blood flow, and oxidative damage collectively contribute to cognitive impairment and neurological dysfunction observed in septic patients. Understanding these mechanisms is crucial for the development of targeted therapeutic strategies aimed at mitigating brain injury and improving outcomes in sepsis.

5. The complex relationship between sepsis and BDNF

5.1 Dysregulation of BDNF in sepsis

Sepsis disrupts the normal regulation of BDNF expression and release in the brain, leading to dysregulation of this crucial neurotrophic factor. Studies have shown that sepsis is associated with reduced BDNF levels in various brain regions, including the hippocampus and cortex [22]. The dysregulation of BDNF in sepsis may result from the systemic inflammatory response, oxidative stress, and alterations in neurotransmitter balance [18, 19].

The release of pro-inflammatory cytokines during sepsis, such as IL-1β and TNF-α, can directly impact BDNF expression. These cytokines have been shown to downregulate BDNF mRNA and protein levels, contributing to the overall decrease in BDNF availability in septic brains [19, 20]. Additionally, the dysregulation of neurotransmitters, particularly glutamate and gamma-aminobutyric acid (GABA), in sepsis can influence BDNF expression, as these neurotransmitters have been shown to modulate BDNF gene transcription.

5.2 Impact of sepsis on BDNF Signaling pathways

BDNF exerts its effects on neurons by binding to its receptor, TrkB, and activating downstream signaling pathways. However, in sepsis, alterations in BDNF signaling pathways have been observed, further contributing to sepsis-associated brain dysfunction [11, 12].

The dysregulation of BDNF-TrkB signaling in sepsis can occur at multiple levels. Sepsis-induced inflammatory mediators, such as IL-1β and TNF-α, can interfere with TrkB receptor activation and downstream signaling cascades, impairing BDNF's neuroprotective effects [12]. Additionally, oxidative stress, which is prevalent in sepsis, can disrupt BDNF signaling pathways, leading to impaired neuroplasticity and synaptic function [23].

Furthermore, sepsis-induced alterations in intracellular signaling pathways, such as the mitogen-activated protein kinase (MAPK) pathway and the phosphoinositide 3-kinase (PI3K)/Akt pathway, can impact BDNF-mediated cellular processes [24]. Dysregulation of these pathways can compromise neuronal survival, synaptic plasticity, and cognitive function, which are key functions influenced by BDNF [24, 25].

5.3 Experimental evidence supporting the involvement of BDNF in sepsis-associated brain dysfunction

Experimental studies have provided compelling evidence for the involvement of BDNF in sepsis-associated brain dysfunction. Animal models of sepsis have demonstrated dysregulation of BDNF expression and signaling in the brain, leading to cognitive impairment and neuronal damage [25, 26].

In these animal models, sepsis-induced neuroinflammation and oxidative stress have been shown to downregulate BDNF expression and impair BDNF signaling pathways. This dysregulation is associated with cognitive deficits, including memory impairment and learning difficulties [27]. Conversely, interventions that enhance BDNF signaling, such as BDNF supplementation or pharmacological agents targeting BDNF pathways, have shown promising results in mitigating cognitive dysfunction and reducing brain injury in septic animals [28].

Furthermore, clinical studies have provided evidence supporting the involvement of BDNF in sepsis-associated brain dysfunction in human patients. Reduced BDNF levels have been observed in the serum and cerebrospinal fluid of septic patients with cognitive impairment compared to those without neurological complications [29, 30]. These studies have also revealed a correlation between lower BDNF levels and worse long-term cognitive outcomes in septic patients.

These findings highlight the potential role of BDNF as a diagnostic and prognostic marker for sepsis-associated brain dysfunction. Furthermore, they suggest that targeting BDNF and modulating its signaling pathways could be a therapeutic approach to mitigate sepsis-induced brain injury and cognitive impairment [30, 31]. Strategies aimed at restoring BDNF levels or enhancing BDNF signaling may hold promise for improving outcomes in septic patients.

In conclusion, the relationship between sepsis and BDNF is complex and multifaceted. Sepsis disrupts the normal regulation of BDNF expression and impairs its signaling pathways in the brain, contributing to sepsis-associated brain dysfunction. The dysregulation of BDNF in sepsis is influenced by factors such as systemic inflammatory response, oxidative stress, and alterations in neurotransmitter balance. Experimental evidence from animal models and clinical studies supports the involvement of BDNF in sepsis-associated cognitive impairment and brain injury.

Understanding the intricate interplay between sepsis and BDNF is crucial for the development of targeted interventions to mitigate sepsis-associated brain dysfunction. Further research is needed to unravel the specific mechanisms by which sepsis dysregulates BDNF and to explore therapeutic strategies aimed at modulating BDNF signaling [31]. By elucidating the role of BDNF in sepsis-associated brain dysfunction, we can pave the way for potential interventions that could improve patient outcomes and reduce the long-term cognitive consequences of sepsis.

Overall, the complex relationship between sepsis and BDNF highlights the importance of investigating the molecular and cellular mechanisms underlying sepsis-associated brain dysfunction. By advancing our understanding of this relationship, we can potentially identify novel therapeutic targets and develop strategies to preserve brain function and improve the quality of life for septic patients.

6. Clinical implications and diagnostic potential

6.1 Biomarker potential of BDNF in sepsis

One of the clinical implications of the relationship between sepsis and BDNF lies in the biomarker potential of BDNF for sepsis diagnosis and prognosis. BDNF levels have been investigated as potential biomarkers to aid in the early detection of sepsis and assess disease severity [32]. Reduced BDNF levels have been observed in septic patients, particularly those with sepsis-associated brain dysfunction. Monitoring BDNF levels could serve as a useful tool in identifying patients at risk of developing cognitive impairment and neurological complications in sepsis.

6.2 BDNF as a predictor of long-term cognitive outcomes in septic patients

The dysregulation of BDNF in sepsis may have implications for long-term cognitive outcomes in septic patients. Clinical studies have shown a correlation between lower BDNF levels and worse cognitive function in septic patients. BDNF could serve as a potential predictor of long-term cognitive impairment and aid in stratifying patients based on their risk of cognitive decline following sepsis [33, 34]. This information could guide post-sepsis management and rehabilitation strategies, allowing for targeted interventions to improve cognitive outcomes.

6.3 BDNF-targeted therapeutic strategies for sepsis-associated brain dysfunction

Given the involvement of BDNF in sepsis-associated brain dysfunction, targeting BDNF and its signaling pathways may offer potential therapeutic strategies. The restoration of BDNF levels or the enhancement of BDNF signaling could help mitigate brain injury and cognitive impairment in septic patients [34].

Pharmacological approaches, such as exogenous BDNF supplementation or pharmacological agents that promote BDNF release or enhance TrkB receptor activation, are being investigated as potential therapeutic interventions. These approaches aim to counteract the dysregulation of BDNF in sepsis and restore its neuroprotective and neuroplasticity-promoting effects [32, 33].

Non-pharmacological interventions, such as physical exercise and environmental enrichment, have also shown promise in upregulating BDNF levels and improving cognitive function in animal models [35]. These interventions may have translational potential for septic patients, as they are relatively safe and accessible therapeutic strategies that could be incorporated into post-sepsis rehabilitation programs.

However, it is important to note that the translation of BDNF-targeted therapeutic strategies into clinical practice requires further research and rigorous clinical trials. The complexity of sepsis and the multifactorial nature of sepsis-associated brain dysfunction necessitate a comprehensive understanding of the mechanisms involved and careful evaluation of potential therapeutic interventions.

In conclusion, BDNF holds clinical implications and diagnostic potential in the context of sepsis-associated brain dysfunction. It has the potential to serve as a biomarker for sepsis diagnosis and prognosis, as well as a predictor of long-term cognitive outcomes in septic patients. Additionally, BDNF-targeted therapeutic strategies, both pharmacological and non-pharmacological, offer promising avenues for mitigating sepsis-induced brain injury and cognitive impairment. Further research and clinical trials are needed to validate the clinical utility of BDNF and to develop effective interventions for improving outcomes in septic patients.

7. Conclusion

7.1 Recap of the complex connection between sepsis and BDNF

The relationship between sepsis and BDNF is multifaceted. Sepsis dysregulates BDNF expression and signaling, contributing to sepsis-associated brain dysfunction. The dysregulation of BDNF is influenced by factors such as the systemic inflammatory response, oxidative stress, and alterations in neurotransmitter balance. Experimental evidence from animal models and clinical studies supports the involvement of BDNF in sepsis-associated cognitive impairment and brain injury.

7.2 Importance of further research in understanding and targeting BDNF in septic patients

To fully harness the potential of BDNF as a diagnostic marker and therapeutic target in septic patients, further research is necessary. Understanding the molecular and cellular mechanisms underlying the dysregulation of BDNF in sepsis is crucial. Additionally, rigorous clinical trials are needed to evaluate the safety and efficacy of BDNF-based interventions in mitigating sepsis-associated brain dysfunction.

7.3 Potential for BDNF-based interventions to mitigate sepsis-associated brain dysfunction and improve patient outcomes

Despite the challenges, BDNF-based interventions hold promise for improving outcomes in septic patients. Strategies aimed at restoring BDNF levels or enhancing

BDNF signaling pathways may help mitigate sepsis-induced brain injury, preserve brain function, and improve cognitive outcomes [35]. The potential diagnostic and prognostic value of BDNF in sepsis further emphasizes the importance of investigating and targeting BDNF in septic patients.

In conclusion, the complex connection between sepsis and BDNF highlights the need for further research and clinical exploration. By unraveling the intricacies of BDNF dysregulation and developing effective interventions, we can potentially improve the diagnosis, prognosis, and management of sepsis-associated brain dysfunction, ultimately leading to better outcomes for septic patients.

Conflict of interest

The authors declare no conflict of interest.

Author details

Ejder Saylav Bora
Izmir Ataturk Research and Training Hospital, Izmir, Türkiye

*Address all correspondence to: saylavbora@hotmail.com

References

[1] Kim J. MD and others, 490. Brain-derived neurotrophic factor (BDNF) is a prognostic biomarker and therapeutic target through AMP-activated protein kinase (AMPK) signal pathway for sepsis. Open Forum Infectious Diseases. 2022;**9**(Supplement_2):ofac492. DOI: 10.1093/ofid/ofac492.548

[2] Tsutsui Y, Yoshio S, Tomiyama T, Shimagaki T, Itoh S, Harada N, et al. Preoperative serum brain-derived neurotrophic factor as a predictive biomarker for sepsis after living-donor liver transplantation. Hepatology Research: the Official Journal of the Japan Society of Hepatology. 2023;**53**(1):72-83. DOI: 10.1111/hepr.13834

[3] Yu M, Qin C, Li P, Zhang Y, Wang Y, Zhang J, et al. Hydrogen gas alleviates sepsis-induced neuroinflammation and cognitive impairment through regulation of DNMT1 and DNMT3a-mediated BDNF promoter IV methylation in mice. International Immunopharmacology. 2021;**95**:107583. DOI: 10.1016/j.intimp.2021.107583

[4] Pisani A, Paciello F, Del Vecchio V, Malesci R, De Corso E, Cantone E, et al. The role of BDNF as a biomarker in cognitive and sensory neurodegeneration. Journal of Personalized Medicine. 2023;**13**(4):652. DOI: 10.3390/jpm13040652

[5] Gotts JE, Matthay MA. Sepsis: Pathophysiology and clinical management. BMJ (Clinical research ed.). 2016;**353**:i1585. DOI: 10.1136/bmj.i1585

[6] Huang M, Cai S, Su J. The pathogenesis of sepsis and potential therapeutic targets. International Journal of Molecular Sciences. 2019;**20**(21):5376. DOI: 10.3390/ijms20215376

[7] Liu D, Huang SY, Sun JH, Zhang HC, Cai QL, Gao C, et al. Sepsis-induced immunosuppression: Mechanisms, diagnosis and current treatment options. Military Medical Research. 2022;**9**(1):56. DOI: 10.1186/s40779-022-00422-y

[8] Mahmoud AA, Abd El Naby SA, Abdelgawad AS, Rizq MS, Abd El Hady NMS. Brain-derived neurotrophic factor and neuroimaging in pediatric patients with sickle cell disease [published online ahead of print, 2023 Feb 11]. Pediatric Research. 2023. DOI: 10.1038/s41390-023-02513-5

[9] Shan D, Zheng Y, Froud K. Brain-derived neurotrophic factor as a clinical biomarker in predicting the development of post-stroke depression: A review of evidence. Cureus. 2021;**13**(6):e15662. DOI: 10.7759/cureus.15662

[10] Kowiański P, Lietzau G, Czuba E, Waśkow M, Steliga A, Moryś J. BDNF: A key factor with multipotent impact on brain Signaling and synaptic plasticity. Cellular and Molecular Neurobiology. 2018;**38**(3):579-593. DOI: 10.1007/s10571-017-0510-4

[11] Yoo JM, Lee BD, Sok DE, Ma JY, Kim MR. Neuroprotective action of N-acetyl serotonin in oxidative stress-induced apoptosis through the activation of both TrkB/CREB/BDNF pathway and Akt/Nrf2/antioxidant enzyme in neuronal cells. Redox Biology. 2017;**11**:592-599. DOI: 10.1016/j.redox.2016.12.034

[12] Zhang JC, Yao W, Hashimoto K. Brain-derived neurotrophic factor (BDNF)-TrkB Signaling in inflammation-related

depression and potential therapeutic targets. Current Neuropharmacology. 2016;**14**(7):721-731. DOI: 10.2174/1570159x14666160119094646

[13] Molnár L, Fülesdi B, Németh N, Molnár C. Sepsis-associated encephalopathy: A review of literature. Neurology India. 2018;**66**(2):352-361. DOI: 10.4103/0028-3886.227299

[14] Tauber SC, Djukic M, Gossner J, Eiffert H, Brück W, Nau R. Sepsis-associated encephalopathy and septic encephalitis: An update. Expert Review of Anti-Infective Therapy. 2021;**19**(2):215-231. DOI: 10.1080/14787210.2020.1812384

[15] Gao Q, Hernandes MS. Sepsis-associated encephalopathy and blood-brain barrier dysfunction. Inflammation. 2021;**44**(6):2143-2150. DOI: 10.1007/s10753-021-01501-3

[16] Madhu LN, Kodali M, Attaluri S, Shuai B, Melissari L, Rao X, et al. Melatonin improves brain function in a model of chronic gulf war illness with modulation of oxidative stress, NLRP3 inflammasomes, and BDNF-ERK-CREB pathway in the hippocampus. Redox Biology. 2021;**43**:101973. DOI: 10.1016/j.redox.2021.101973

[17] Netto MB, de Oliveira Junior AN, Goldim M, Mathias K, Fileti ME, da Rosa N, et al. Oxidative stress and mitochondrial dysfunction contributes to postoperative cognitive dysfunction in elderly rats. Brain, Behavior, and Immunity. 2018;**73**:661-669. DOI: 10.1016/j.bbi.2018.07.016

[18] Bouvier E, Brouillard F, Molet J, Claverie D, Cabungcal JH, Cresto N, et al. Nrf2-dependent persistent oxidative stress results in stress-induced vulnerability to depression. Molecular Psychiatry. 2017;**22**(12):1701-1713. DOI: 10.1038/mp.2016.144

[19] Garbossa L, Joaquim L, Danielski LG, Goldim MPS, Machado RS, Metzker K, et al. The effect of modafinil on passive avoidance memory, brain level of BDNF and oxidative stress markers in sepsis survivor rats. The International Journal of Neuroscience. 2022:1-9. Advance online publication. DOI: 10.1080/00207454.2022.2154076

[20] Jiang S, Bai L, Zhang X, Zhou X, Liu Y. Preexposure to heat stress attenuates sepsis-associated inflammation and cognitive decline in rats. Neuroscience Letters. 2022;**780**:136647. DOI: 10.1016/j.neulet.2022.136647

[21] de Souza Stork S, Hübner M, Biehl E, Danielski LG, Bonfante S, Joaquim L, et al. Diabetes exacerbates sepsis-induced Neuroinflammation and brain mitochondrial dysfunction. Inflammation. 2022;**45**(6):2352-2367. DOI: 10.1007/s10753-022-01697-y

[22] Giustina AD, de Souza Goldim MP, Danielski LG, Garbossa L, Junior ANO, Cidreira T, et al. Lipoic acid and fish oil combination potentiates Neuroinflammation and oxidative stress regulation and prevents cognitive decline of rats after sepsis. Molecular Neurobiology. 2020;**57**(11):4451-4466. DOI: 10.1007/s12035-020-02032-y

[23] Liu K, Wan G, Jiang R, Zou L, Wan D, Zhu H, et al. Astragalus injection ameliorates lipopolysaccharide-induced cognitive decline via relieving acute neuroinflammation and BBB damage and upregulating the BDNF-CREB pathway in mice. Pharmaceutical Biology. 2022;**60**(1):825-839. DOI: 10.1080/13880209.2022.2062005

[24] Gürkan G, Atasoy Ö, Çini N, et al. Reparative, neuroprotective and

anti-neurodegenerative effects of granulocyte colony stimulating factor in radiation-induced brain injury model [published online ahead of print, 2023 May 11]. Journal of Korean Neurosurgical Society. 2023. DOI: 10.3340/jkns.2023.0049

[25] Jin Y, Sun LH, Yang W, Cui RJ, Xu SB. The role of BDNF in the Neuroimmune Axis regulation of mood disorders. Frontiers in Neurology. 2019;**10**:515. DOI: 10.3389/fneur.2019.00515

[26] Zhao XP, Li H, Dai RP. Neuroimmune crosstalk through brain-derived neurotrophic factor and its precursor pro-BDNF: New insights into mood disorders. World Journal of Psychiatry. 2022;**12**(3):379-392. DOI: 10.5498/wjp.v12.i3.379

[27] Yap NY, Toh YL, Tan CJ, Acharya MM, Chan A. Relationship between cytokines and brain-derived neurotrophic factor (BDNF) in trajectories of cancer-related cognitive impairment. Cytokine. 2021;**144**:155556. DOI: 10.1016/j.cyto.2021.155556

[28] Mahalakshmi B, Maurya N, Lee SD, Bharath Kumar V. Possible neuroprotective mechanisms of physical exercise in neurodegeneration. International Journal of Molecular Sciences. 2020;**21**(16):5895. DOI: 10.3390/ijms21165895

[29] Carniel BP, da Rocha NS. Brain-derived neurotrophic factor (BDNF) and inflammatory markers: Perspectives for the management of depression. Progress in Neuro-Psychopharmacology and Biological Psychiatry. 2021;**108**:110151. DOI: 10.1016/j.pnpbp.2020.110151

[30] Xie J, Zhao ZZ, Li P, Zhu CL, Guo Y, Wang J, et al. Senkyunolide I protects against sepsis-associated encephalopathy by attenuating sleep deprivation in a murine model of Cecal ligation and puncture. Oxidative Medicine and Cellular Longevity. 2021;**2021**:6647258. DOI: 10.1155/2021/6647258

[31] Bai Y, Han Q, Dong B, Lin H, Jiang Y, Zhang X, et al. PPARα contributes to the therapeutic effect of hydrogen gas against sepsis-associated encephalopathy with the regulation to the CREB-BDNF signaling pathway and hippocampal neuron plasticity-related gene expression. Brain Research Bulletin. 2022;**184**:56-67. DOI: 10.1016/j.brainresbull.2022.03.015

[32] Ji MH, Tang H, Luo D, Qiu LL, Jia M, Yuan HM, et al. Environmental conditions differentially affect neurobehavioral outcomes in a mouse model of sepsis-associated encephalopathy. Oncotarget. 2017;**8**(47):82376-82389. DOI: 10.18632/oncotarget.19595

[33] Murawska-Ciałowicz E, Wiatr M, Ciałowicz M, Gomes de Assis G, Borowicz W, Rocha-Rodrigues S, et al. BDNF impact on biological markers of depression-role of physical exercise and training. International Journal of Environmental Research and Public Health. 2021;**18**(14):7553. DOI: 10.3390/ijerph18147553

[34] Gao L, Zhang Y, Sterling K, Song W. Brain-derived neurotrophic factor in Alzheimer's disease and its pharmaceutical potential. Translational Neurodegeneration. 2022;**11**(1):4. DOI: 10.1186/s40035-022-00279-0

[35] Gaitán JM, Moon HY, Stremlau M, Dubal DB, Cook DB, Okonkwo OC, et al. Effects of aerobic exercise training on systemic biomarkers and cognition in late middle-aged adults at risk for Alzheimer's disease. Frontiers in Endocrinology. 2021;**12**:660181. DOI: 10.3389/fendo.2021.660181

Section 4

The Effect of Exercise and Vitamin D on Brain-Derived Neurotrophic Factor

Chapter 6

Combined Exercise and Vitamin D on Brain-Derived Neurotrophic Factor

Rastegar Hoseini, Zahra Hoseini and Elahe Bahmani

Abstract

Brain-derived neurotrophic factor (BDNF) is a highly conserved neurotrophic protein of the nerve growth factor family. Neurotrophins are proteins that help to stimulate and control neurogenesis, BDNF being the most active one. BDNF may be useful in the prevention and management of several diseases including Multiple Sclerosis (MS) and Diabetes. Lifestyle modifications (physical activity and diet) are among the most promising strategies for altering BDNF levels. In this chapter, we aimed to investigate the effects of aerobic and resistance training and combined exercise and vitamin D therapy on BDNF levels.

Keywords: aerobic training, resistance training, vitamin D, health, BDNF

1. Introduction

Brain-derived neurotrophic factor (BDNF) is a protein that belongs to the nerve growth factor (NGF) family and has been conserved throughout evolution. It plays a significant role in regulating synapses, affecting both their structure and function in multiple areas of the brain. BDNF also helps promote neuron survival, neuroplasticity, neurite growth, and synaptogenesis. BDNF is an important factor affecting cognitive function which has recently interested a bulk trend of effort in the health context [1–3]. BDNF was first isolated from the pig brain in 1982 by Yves-Alain Barde and Hans Thoenen [4] which was then cloned in 1989 [5]. BDNF is one of the neurotrophic factors that support differentiation. BDNF is a protein that, in humans, is encoded by the BDNF gene [6]. BDNF is one of the neurotrophic factors that support the differentiation, maturation, and survival of neurons in the nervous system and shows a neuroprotective effect under adverse conditions, such as glutamatergic stimulation, cerebral ischemia, hypoglycemia, and neurotoxicity [7–9]. BDNF is a member of the neurotrophins family of growth factors, which are related to the canonical NGF, a family which also includes NT-3 and NT-4/NT-5. It is widely expressed in the CNS [10], retina, kidneys, prostate, motor neurons, and skeletal muscle and is also found in saliva (31). BDNF binds to its high-affinity cell surface receptors, tyrosine kinase B (TrkB), and activates signal transduction cascades (IRS1/2, PI3K, Akt) [11], crucial for CREB and CBP production, that encode proteins involved in β-cell survival [12]. TrkB are part of the larger family of protein tyrosine kinases, encompassing the receptor tyrosine kinase proteins which contain a transmembrane domain, as well as the non-receptor tyrosine kinases which do not possess transmembrane domains [13]. Of the 90 unique tyrosine kinase genes identified in the human genome, 58 encode TrkB [14]. TrkB has a crucial function in both regular cell processes and the advancement of various cancer types [15]. According to research, mutations in TrkB result in the activation of signaling pathways that influence protein expression [16]. BDNF and insulin-like growth factor-1 have similar downstream signaling mechanisms incorporating both p-CAMK and MAPK that increase the expression of pro-survival genes [17]. BDNF protein and mRNA have been identified in most brain areas including the olfactory bulb, cortex, hippocampus, basal forebrain, mesencephalon, hypothalamus, brainstem, and spinal cord [17, 18] which stimulates and controls neurogenesis which is the growth of new neurons from neural stem cells [19]. Decreased levels of BDNF are associated with neurodegenerative diseases with neuronal loss, such as Parkinson's disease (15), Alzheimer's disease (27), Multiple Sclerosis (MS) (16), and Huntington's disease (17). Besides the neuroprotective effect, BDNF plays a major role in energy homeostasis. The peripheral or intracerebroventricular (ICV) BDNF administration suppresses energy intake and reduces body weight [20]. BDNF has been identified as a key component of the hypothalamic signaling pathway. This explains why BDNF controls body weight, decreases food intake, lowers blood glucose levels and controls energy homeostasis [20].

BDNF plays an important role in neuronal survival and growth, serves as a neurotransmitter modulator, and participates in neuronal plasticity, which is essential for learning and memory [21]. BDNF is responsible for making your neurons stronger. Nevertheless, BDNF isoforms have also been observed to affect neuronal activity by being associated with cellular models of memory (i.e., long-term potentiation and long-term depression) [22]. Neurotrophic factors regulate neuronal differentiation, phenotype maintenance, and synaptic sprouting [23]. They also protect adult neurons from mechanical, toxic, or ischemic injuries and interfere with the death of neurons by necrosis or apoptosis [24]. Lifestyle modifications (physical activity and diet) are among the most promising strategies for altering BDNF levels. We aimed to investigate the effects of aerobic and resistance training and combined exercise and vitamin D therapy on BDNF levels.

2. Function of brain-derived neurotrophic factor (BDNF)

BDNF may be useful in the prevention and management of several diseases including MS and Diabetes [25, 26]. In the brain, it is active in the hippocampus, cortex, and basal forebrain areas vital to learning, memory, and higher thinking [27]. Although the vast majority of neurons in the mammalian brain are formed prenatally, parts of the adult brain retain the ability to grow new neurons from neural stem cells in a process known as neurogenesis. BDNF acts on certain neurons of the central nervous system and the peripheral nervous system expressing TrkB, helping to support the survival of existing neurons, and encouraging growth and differentiation of new neurons and synapses [28]. Neurotrophins are proteins that help to stimulate and control neurogenesis, BDNF being one of the most active ones [29]. Endogenous BDNF is known to be involved in cellular development and growth, mood regulation, and cognitive functions such as learning and memory. BDNF appears to be a crucial regulatory mechanism in the growth and development of neurons across various regions of the brain. It has also been shown to enhance neuron survival by increasing resistance to nerve damage [30]. Mice born without the ability to make BDNF have developmental defects in the brain and sensory nervous system, and usually die soon after birth suggesting that BDNF plays an important role in normal neural development. BDNF also regulates both excitatory and inhibitory synaptic transmission and activity-dependent plasticity as a key molecule involved in plastic changes related to learning and memory [31]. Emerging data indicate that the induction of localized axonal synthesis by BDNF underlies its role in regulating synaptic efficacy. Changes in BDNF expression are associated with both normal and pathological aging and also psychiatric disease, in particular in structures important for memory processes such as the hippocampus and para-hippocampal areas; as a results, BDNF itself is important for long-term memory [32]. BDNF has a role in axonal guidance and regulates activity-dependent synaptic plasticity and long-term potentiation [33]. Neurotrophins are essential for short-term neuronal plasticity and long-term neuroprotection in the CNS. The survival and morphogenesis of CNS neurons depend on BDNF/TrkB-stimulated signaling. Activation of different intracellular signaling pathways, including MAPK/ERK, PLCγ, and PI3K, is triggered when BDNF binds to TrkB. These mechanisms are responsible for the biological effects that BDNF has on neurons [34]. BDNF/TrkB-stimulated intracellular signaling is critical for neuronal survival, morphogenesis, and plasticity [35]. BDNF regulates glucose and energy metabolism and prevents the exhaustion of β cells [36]. Findings also indicate that BDNF is involved in both central metabolic pathways and the mediation of energy metabolism in peripheral organs. Recent findings suggest that the BDNF signaling pathway in the hypothalamus may have the ability to regulate energy balance, control body weight, and influence feeding behavior [37]. BDNF is a protein produced by muscle cells during exercise that can enhance the breakdown of fat in skeletal muscles through a process dependent on AMP-activated protein kinase [38].

3. The effect of aerobic exercise on the brain-derived neurotrophic factor (BDNF)

Low circulating BDNF levels have been associated with a wide range of neuropsychiatric disorders including depression, schizophrenia, and neurodegenerative diseases, although no causal relationship has yet been established [39, 40]. Over the last 10 years, studies have examined what causes short-term and long-term increases in BDNF levels in animal brains and human blood. These studies assume that higher levels of BDNF can benefit brain health [41]. Certain types of physical exercise have been shown to markedly (threefold) increase BDNF synthesis in the human brain, a phenomenon that is partly responsible for exercise-induced neurogenesis and improvements in cognitive function [42]. The release of BNDF in humans is stimulated by physical activity and may be related to improvements in executive function [43]. Executive function is responsible for higher cognitive processes involved in managing other basic cognitive functions. Aerobic exercise is proposed to induce the expression of BDNF throughout the central nervous system, which in turn, can enhance synaptic plasticity (52). Research has consistently shown that aerobic exercise can elevate baseline BDNF levels in the hippocampus, striatum, and various cortical regions in laboratory animals [44]. Encouragingly, BDNF transcription can be induced in the rat hippocampus after only three consecutive days of aerobic exercise [45]. Exercise promotes the expression of BDNF through the action of the ketone body β-hydroxybutyrate [46]. A form of physical activity known as aerobic exercise has been proven to have positive effects on individuals with neurological disorders who undergo this type of training [47]. For example, after a program of aerobic exercise, individuals with stroke [48], MS (16), and Parkinson's disease (15) have shown improvements in walking, functional ability, and motor performance. In addition to gains in cardiorespiratory fitness, exercise-induced increases in BDNF levels in the motor cortex and hippocampus have also been associated with enhanced learning and memory [49]. Additionally, it guides decision-making processes for motor tasks and healthy behaviors. It has been suggested that this increase in BDNF is associated with enhanced hippocampal synaptic plasticity, which supposedly enhances synaptic transmission and increases the expression of molecules associated with learning and memory [50]. Recent research indicates that the levels of BDNF, which increase after short-term exercise, can continue to rise with long-term aerobic exercise [51]. Long-term endurance training in humans has been shown to result in an increase in resting serum BDNF levels that persist over time [52]. In contrast, some studies have reported that the duration of aerobic exercise does not have a significant influence on resting levels of serum BDNF [51, 53]. Several mechanisms have been proposed to explain the positive impacts of aerobic exercise. These include increased cerebral blood flow, changes in neurotransmitter release, structural changes in the central nervous system, and altered arousal levels [54]. Serotonin levels regulate BDNF, which is a potential cause of serotonin-delivering axon growth. Similar to exercise, antidepressants increase BDNF levels, which could explain their effectiveness in improving mood [55]. A recent review and meta-analysis of 29 studies investigating the effect of exercise on BDNF in healthy humans found that a single session of aerobic exercise significantly increases BDNF levels immediately post-exercise demonstrating a moderate effect [56]. Furthermore, in the same review, a program of aerobic training was shown to significantly increase resting levels of BDNF, with a small effect size [56]. These findings provide evidence that both single and long-term aerobic exercise has a significant impact on BDNF levels in healthy humans.

Aerobic training has been shown to improve brain function, and one of the mechanisms behind this effect is thought to be an increase in BDNF [57]. BDNF is a protein that promotes the growth and survival of neurons in the brain, and it plays a key role in learning, memory, and cognitive function [58]. Studies have found that aerobic exercise can increase levels of BDNF in both animals and humans [59]. This may be because exercise stimulates the release of various growth factors, including BDNF [60]. In addition, exercise has been shown to increase blood flow to the brain, which may also contribute to the increase in BDNF levels [60]. Once released into the brain, BDNF binds to specific receptors on neurons and triggers a cascade of molecular events that promote neuron survival and growth. These events include the activation of various signaling pathways, such as the MAPK/ERK pathway, which leads to increased protein synthesis and enhanced neuronal plasticity. Overall, these cellular and molecular mechanisms suggest that aerobic exercise can have a powerful impact on brain function by increasing levels of BDNF [61]. By promoting neuron survival and growth, BDNF may help support cognitive function and protect against age-related decline [62].

4. The effect of resistance exercise on the brain-derived neurotrophic factor (BDNF)

Strength training is a staple for physical and mental health. The benefits are not only stronger bones, ligaments, tendons, and muscle tissues but also a more capable mind [63]. Recently studies demonstrated that resistance exercise can also elevate BDNF levels in the hippocampus (50). Exercise has been proven to promote neurogenesis by increasing BDNF and lowering cortisol [64]. The exact process and mechanism by which resistance exercise increases BDNF, leading to changes in neuroplasticity, is not yet fully understood (41). There were contradictory results in the literature regarding the response of BDNF to resistance training some found positive BDNF response while some reported no differences between BDNF levels before and after training. Lodo et al. investigated the response of neurotrophic factors in schemes of equal volume consisting of two resistance training sessions with 1 week of rest between the sessions with a total of 30 participants suggesting that the intensity of resistance training is not a significant factor in the neurotrophic factor response when the total load lifted is equated in the range of submaximal repetition [65]. Another study investigated the effect of a resistance exercise 3×/week for 6 weeks in two groups of 80% one repetition maximum (1RM) with low repetition and 65% 1RM with high repetition in men with at least 2 years of resistance training experience hypothesizing that a minimum volume and greater proximity to one repetition maximum may be required to elicit a BDNF response [66].

Additionally, studies investigated the effect of short and longer training sessions, only two or three compared to 15–40 sessions, showing significant differences concerning BDNF between these studies, but it is not possible to conclude which (single sessions vs. several sessions) may produce a better BDNF response. The available research on BDNF and its relationship with resistance and strength training yields inconclusive results. From the studies conducted, it appears that high-intensity workouts at 70% or above based on 1RM, low repetition, and specific rest periods are necessary to induce changes in BDNF levels. Additionally, whole-body training or lower-body training with free weights and multi-joint movements may produce more favorable outcomes [67]. Further studies are needed to draw a better conclusion for BDNF response to resistance training.

Resistance training has been shown to have a significant impact on BDNF [68]. This protein is responsible for the growth and survival of neurons in the brain, as well as synaptic plasticity [69]. Resistance training stimulates the production of BDNF through a variety of cellular and molecular mechanisms [70]. One mechanism by which resistance training increases BDNF levels is through the activation of the mTOR pathway. This pathway plays a critical role in regulating protein synthesis and cell growth and has been linked to increased BDNF expression [70]. In addition, resistance training has been shown to increase the activity of AMPK, an enzyme that regulates energy metabolism and promotes mitochondrial biogenesis. This process may also contribute to the upregulation of BDNF. Another cellular mechanism by which resistance training affects BDNF is through modulation of oxidative stress [71]. Exercise-induced oxidative stress can stimulate the expression of antioxidant enzymes, which protect against damage caused by free radicals. These enzymes may also indirectly increase BDNF levels by reducing inflammation and improving overall neuronal health. Finally, resistance training may promote BDNF expression through its effects on neurotransmitter systems [72]. Exercise has been shown to increase dopamine and serotonin release in the brain, both of which are known to stimulate BDNF production. Additionally, exercise-induced changes in glutamate receptor activity may also contribute to increased BDNF expression [73]. In conclusion, resistance training has a profound impact on BDNF levels through a complex interplay of cellular and molecular mechanisms. By promoting the growth and survival of neurons in the brain, resistance training has the potential to enhance cognitive function and improve overall neurological health.

5. The effect of vitamin D on brain-derived neurotrophic factor (BDNF)

Vitamin D is a steroid hormone essential for maintaining calcium metabolism and various extra-skeletal functions. Noteworthy, vitamin D controls more than 1000 genes, including those responsible for the regulation of cellular proliferation, differentiation, apoptosis, and angiogenesis [74]. This steroid hormone plays an important role in the nervous system, including differentiation, calcium regulation, homeostasis, modulation, the release of neurotrophins, and the activity of brain genes and neurotransmitter metabolism enzymes [75, 76]. As a result of restricted sunlight exposure and/or dietary intake, many people are vitamin D deficient and need vitamin D supplements to meet their vitamin D requirement. Very frequently vitamin D insufficiency can occur with several potential consequences, many of which are still under investigation. Vitamin D supplementation acts to improve performance speed and proximal muscle strength, thus reducing the risk of falls, osteoporosis, and fractures in post-menopausal women [77]. In addition to its well-established action in calcium homeostasis, vitamin D is being reconsidered as a neuroprotective steroid. The reported neuroprotective effects of vitamin D include the *in vitro* biosynthesis of neurotrophic factors, the inhibition of nitric oxide synthase, and the increased glutathione levels in the brain detoxification pathways [75]. Vitamin D is potent *in vitro* inducer of NGF mRNA expression in neural brain cells and protects the brain cortex against amyloid-beta-induced toxicity. While, suppression of vitamin D receptor (VDR) in neuronal cultures disrupts L-type voltage-sensitive calcium channels and NGF production, increasing vulnerability to aging and neurodegeneration. Vitamin D and its analogs can cross the blood-brain barrier, and it has been shown that VDR and enzymes involved in the bioactivation and catabolism of vitamin D are abundantly expressed in the brain neural cells, particularly in areas affected by neurodegenerative disorders [23]. Indeed, vitamin D stimulates the expression of P75NTR, the neurotrophins low-affinity receptor. Vitamin D supplementation plays a crucial role in the modulation of neurotrophic factors that may reflect a compensatory mechanism. Confirming animal studies, low levels of circulating vitamin D may cause cognitive decline not affected by neurological impairment in human subjects and this can be reversed by vitamin D substitution therapy [78]. It must be taken into account that postmenopausal women or MS patients, as well as amenorrhoeic subjects, showed lower plasma BDNF levels. Vitamin D and some metabolically active precursors modulate the synthesis of neurotrophins, thus, neurons could therefore be vulnerable to aging and neurodegeneration when there is a long-term or permanent deficiency [79]. Vitamin D's significance in calcium metabolism and neurotrophic factors regulation is crucial to the brain's functioning, as well as BDNF's role in supporting neuron survival. Further research should investigate how vitamin D impacts BDNF-related health outcomes.

One proposed mechanism by which vitamin D may affect BDNF is through its ability to regulate gene expression. Vitamin D receptors are found throughout the body, including in the brain, where they can bind to specific DNA sequences and influence the expression of genes involved in neuroplasticity and cognition [80]. In addition to its effects on gene expression, vitamin D may also modulate BDNF levels through its anti-inflammatory properties [80]. Chronic inflammation has been linked to decreased BDNF levels, and studies have shown that vitamin D can reduce inflammation in both the peripheral and central nervous systems [81]. Finally, research has suggested that vitamin D may interact with other molecules implicated in BDNF regulation, such as serotonin and dopamine. These neurotransmitters play important roles in mood regulation and cognitive function, and their interaction with vitamin D may provide further insight into the mechanisms underlying the relationship between vitamin D and BDNF.

6. The effect of combined exercise & vitamin D on the brain-derived neurotrophic factor (BDNF)

The evidence on the interactive effects of exercise and vitamin D supplementation on neurotrophic factors and neuronal growth is limited and controversial. Exercise has been reported to exert neuroprotection by neurogenesis and angiogenesis. On the one hand, exercise increases growth factor signaling by reducing inflammatory factors and improving growth factor levels [82]. On the other hand, one of the non-invasive treatment approaches proposed in diseases is the use of vitamin D. The findings show the superiority of using combined exercise and vitamin D strategy over exercise or vitamin D alone in increasing BDNF [83]. In addition, the antioxidant effects of vitamin D supplementation and exercise have a positive role in the regulation of neurotrophic factors and growth cells of the nervous system, as well as the function of immune system regulatory cells, due to their direct effect on the secretion of stress-related hormones, including glucocorticoids by reducing the level of oxidative stress and inflammation. In this regard, Horn et al. reported that the increase in BDNF expression following exercise is regulated by neurotransmitters (glutamate, acetylcholine, and serotonin) and GABA receptors and environmental hormones (estrogen, progesterone, and testosterone, growth, and glucocorticoid) [84]. Also, Bahmani et al. reported that combined aerobic training and vitamin D supplementation increased BDNF and NGF, and downregulated CRP, TNF-a, IL-6, and IL-1β more effectively than either alone in MS patients suggesting combined therapy as a better approach to improve neurotrophins and inflammatory biomarker levels in female MS patients [85]. Babaei et al. studied the beneficial effects of aerobic exercise on metabolic syndrome components, cognitive performance, BDNF, and irisin in ovariectomized rats with different serum vitamin D levels reporting that vitamin D insufficiency deteriorates metabolic syndrome components and elevates serum BDNF as a compensatory metabotropic factor, and further high dose of vitamin D supplementation along with aerobic exercise significantly attenuates these components parallel with a reduction in BDNF [86]. Also, another study investigated the effect of aerobic training and vitamin D supplementation on fatigue and quality of life in patients with MS during the COVID-19 outbreak that showed aerobic training and vitamin D supplementation effectively reduced fatigue and improved the QoL in female MS patients in favor of combined protocols than separate protocols [86].

Physical exercise and vitamin D have both been linked to increased levels of BDNF, a protein that plays an important role in the growth and survival of neurons [87]. Research has shown that combining exercise with vitamin D supplementation can lead to even greater increases in BDNF levels. One possible mechanism for this effect is through the regulation of gene expression [88]. Exercise and vitamin D have both been shown to regulate the expression of genes related to BDNF, leading

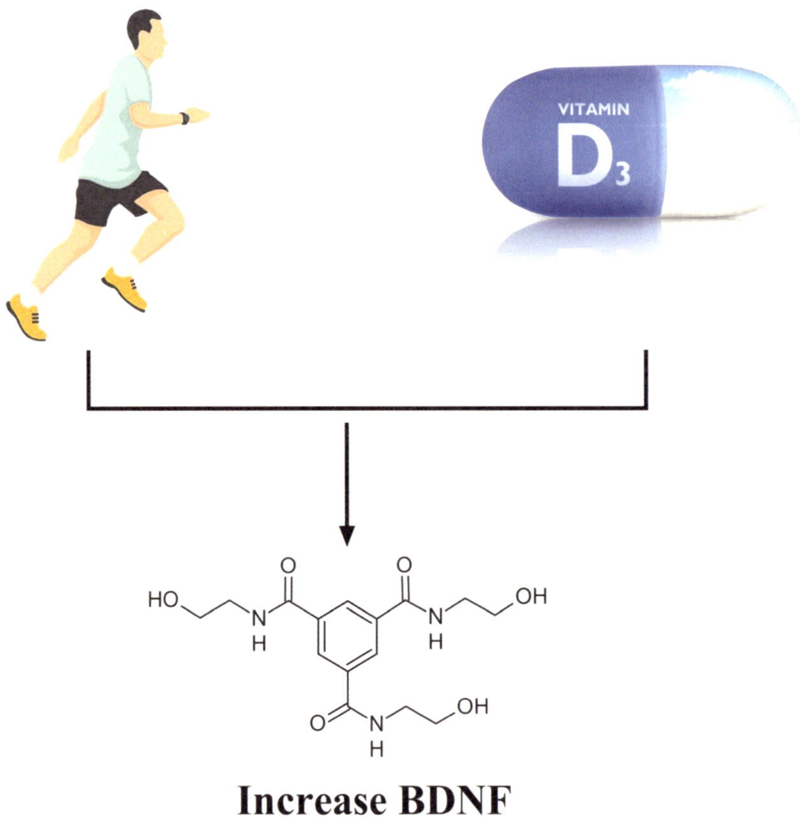

Figure 1.
The combined effect of exercise and vitamin D on BDNF.

to increased production of the protein. Another possible mechanism is through the modulation of inflammation. Both exercise and vitamin D have anti-inflammatory effects, and chronic inflammation has been linked to decreased BDNF levels. By reducing inflammation, exercise and vitamin D may help to increase BDNF production [89]. Exercise also increases blood flow to the brain, which may contribute to the increase in BDNF levels seen with combined exercise and vitamin D supplementation. This increased blood flow may also improve oxygen delivery to neurons, further supporting their survival and growth [90]. Vitamin D has also been shown to play a role in calcium signaling within neurons, which is important for their function and survival [91]. Combined with exercise-induced increases in calcium signaling, this may lead to greater BDNF production. Finally, both exercise and vitamin D have been linked to improvements in mood and cognitive function. These improvements may be mediated by increased BDNF levels, as the protein is known to promote neuronal plasticity and support learning and memory processes [90]. In conclusion, several cellular and molecular mechanisms may explain the beneficial effects of combined exercise and vitamin D on BDNF levels. These mechanisms include regulation of gene expression, reduction of inflammation, increased blood flow to the brain, modulation of calcium signaling, and improvements in mood and cognitive function. Further research is needed to fully understand these mechanisms and how they contribute to overall brain health (**Figure 1**).

DOI: http://dx.doi.org/10.5772/intechopen.112021

7. Conclusions

In conclusion, the combination of exercise and vitamin D has been shown to have a positive effect on BDNF levels. Studies suggest that regular physical activity can increase BDNF levels, while vitamin D supplementation may enhance the effects of exercise on BDNF. These findings have important implications for individuals looking to improve their cognitive function and overall brain health. Incorporating both exercise and sufficient vitamin D intake into one's lifestyle may provide a simple yet effective way to support healthy brain function throughout life. However, more research is needed to fully understand the mechanisms behind these effects and how they vary in different populations.

Conflict of interest

No potential conflict of interest was reported by the authors.

Author details

Rastegar Hoseini*, Zahra Hoseini and Elahe Bahmani
Faculty of Sport Sciences, Department of Exercise Physiology, Razi University, Kermanshah, Iran

*Address all correspondence to: rastegar.hoseini@gmail.com

References

[1] Sun J, Ke Z, Yip SP, Hu X-L, Zheng X-X, Tong K-Y. Gradually increased training intensity benefits rehabilitation outcome after stroke by BDNF upregulation and stress suppression. BioMed Research International. 2014;**20**(1):12-18

[2] Huang Y, Huang C, Yun W. Peripheral BDNF/TrkB protein expression is decreased in Parkinson's disease but not in essential tremor. Journal of Clinical Neuroscience. 2019;**63**:176-181

[3] Du X, Chen W, Zhan N, Bian X, Yu W. The effects of low-intensity resistance training with or without blood flow restriction on serum BDNF, VEGF and perception in patients with post-stroke depression. Neuroendocrinology Letters. 2021;**42**(4):229-235

[4] Knaepen K, Goekint M, Heyman EM, Meeusen R. Neuroplasticity—exercise-induced response of peripheral brain-derived neurotrophic factor: A systematic review of experimental studies in human subjects. Sports Medicine. 2010;**40**:765-801

[5] Gold SM, Schulz K-H, Hartmann S, Mladek M, Lang UE, Hellweg R, et al. Basal serum levels and reactivity of nerve growth factor and brain-derived neurotrophic factor to standardized acute exercise in multiple sclerosis and controls. Journal of Neuroimmunology. 2003;**138**(1-2):99-105

[6] Ferrer A, Labad J, Salvat-Pujol N, Barrachina M, Costas J, Urretavizcaya M, et al. BDNF genetic variants and methylation: Effects on cognition in major depressive disorder. Translational Psychiatry. 2019;**9**(1):265

[7] Ahuja P, Ng CF, Pang BPS, Chan WS, Tse MCL, Bi X, et al. Muscle-generated BDNF (brain derived neurotrophic factor) maintains mitochondrial quality control in female mice. Autophagy. 2022;**18**(6):1367-1384

[8] Li X, Jiang T, Sun X, Yong X, Ma X, Liu J. The relationship between occupational stress, musculoskeletal disorders and the mental health of coal miners: The interaction between BDNF gene, TPH2 gene polymorphism and the environment. Journal of Psychiatric Research. 2021;**135**:76-85

[9] Li C, Meng F, Lei Y, Liu J, Liu J, Zhang J, et al. Leptin regulates exon-specific transcription of the Bdnf gene via epigenetic modifications mediated by an AKT/p300 HAT cascade. Molecular Psychiatry. 2021;**26**(8):3701-3722

[10] Fukuchi M, Okuno Y, Nakayama H, Nakano A, Mori H, Mitazaki S, et al. Screening inducers of neuronal BDNF gene transcription using primary cortical cell cultures from BDNF-luciferase transgenic mice. Scientific Reports. 2019;**9**(1):11833

[11] Jiménez-Treviño L, Saiz PA, García-Portilla MP, Blasco-Fontecilla H, Carli V, Iosue M, et al. 5-HTTLPR–brain-derived neurotrophic factor (BDNF) gene interactions and early adverse life events effect on impulsivity in suicide attempters. The World Journal of Biological Psychiatry. 2019;**20**(2):137-149

[12] Dalir T, Gharakhanlou R, Peeri M, Homaee HM. The effect of four weeks of aerobic training on the expression of Sirt1, CREB and BDNF genes in hippocampus of male wistar rats with alzheimer's disease. Journal of Ardabil University of Medical Sciences. 2020;**20**(4):562-574

[13] Sartor GC, Malvezzi AM, Kumar A, Andrade NS, Wiedner HJ, Vilca SJ, et al.

Enhancement of BDNF expression and memory by HDAC inhibition requires BET bromodomain reader proteins. Journal of Neuroscience. 2019;**39**(4):612-626

[14] Ying X, Tu W, Li S, Wu Q, Chen X, Zhou Y, et al. Hyperbaric oxygen therapy reduces apoptosis and dendritic/synaptic degeneration via the BDNF/TrkB signaling pathways in SCI rats. Life Sciences. 2019;**229**:187-199

[15] Mondal AC, Fatima M. Direct and indirect evidences of BDNF and NGF as key modulators in depression: Role of antidepressants treatment. International Journal of Neuroscience. 2019;**129**(3):283-296

[16] De Vincenti AP, Ríos AS, Paratcha G, Ledda F. Mechanisms that modulate and diversify BDNF functions: Implications for hippocampal synaptic plasticity. Frontiers in Cellular Neuroscience. 2019;**13**:135

[17] Pradhan J, Noakes PG, Bellingham MC. The role of altered BDNF/TrkB signaling in amyotrophic lateral sclerosis. Frontiers in Cellular Neuroscience. 2019;**13**:368

[18] Shen T, You Y, Joseph C, Mirzaei M, Klistorner A, Graham SL, et al. BDNF polymorphism: A review of its diagnostic and clinical relevance in neurodegenerative disorders. Aging and Disease. 2018;**9**(3):523

[19] Zhang J-C, Yao W, Hashimoto K. Brain-derived neurotrophic factor (BDNF)-TrkB signaling in inflammation-related depression and potential therapeutic targets. Current Neuropharmacology. 2016;**14**(7):721-731

[20] Vanevski F, Xu B. Molecular and neural bases underlying roles of BDNF in the control of body weight. Frontiers in Neuroscience. 2013;7:37

[21] El Hayek L, Khalifeh M, Zibara V, Abi Assaad R, Emmanuel N, Karnib N, et al. Lactate mediates the effects of exercise on learning and memory through SIRT1-dependent activation of hippocampal brain-derived neurotrophic factor (BDNF). Journal of Neuroscience. 2019;**39**(13):2369-2382

[22] Roustazade R, Radahmadi M, Yazdani Y. Therapeutic effects of saffron extract on different memory types, anxiety, and hippocampal BDNF and TNF-α gene expressions in sub-chronically stressed rats. Nutritional Neuroscience. 2022;**25**(1):192-206

[23] Yurekli U, Tunc Z. Correlation between vitamin D, homocysteine and brain-derived neurotrophic factor levels in patients with ischemic stroke. European Review for Medical and Pharmacological Sciences. 2022;**26**(21):8004-8010

[24] Wang Q, Liu J, Guo Y, Dong G, Zou W, Chen Z. Association between BDNF G196A (Val66Met) polymorphism and cognitive impairment in patients with Parkinson's disease: A meta-analysis. Brazilian Journal of Medical and Biological Research. 2019;**52**(8):1-8

[25] Ng TKS, Ho CSH, Tam WWS, Kua EH, Ho RC-M. Decreased serum brain-derived neurotrophic factor (BDNF) levels in patients with Alzheimer's disease (AD): A systematic review and meta-analysis. International Journal of Molecular Sciences. 2019;**20**(2):257

[26] Rozanska O, Uruska A, Zozulinska-Ziolkiewicz D. Brain-derived neurotrophic factor and diabetes. International Journal of Molecular Sciences. 2020;**21**(3):841

[27] Stadelmann C, Kerschensteiner M, Misgeld T, BruÈck W, Hohlfeld R,

Lassmann H. BDNF and gp145trkB in multiple sclerosis brain lesions: Neuroprotective interactions between immune and neuronal cells? Brain. 2002;**125**(1):75-85

[28] Lin TW, Harward SC, Huang YZ, McNamara JO. Targeting BDNF/TrkB pathways for preventing or suppressing epilepsy. Neuropharmacology. 2020;**167**:107734

[29] Islam R, Matsuzaki K, Sumiyoshi E, Hossain ME, Hashimoto M, Katakura M, et al. Theobromine improves working memory by activating the CaMKII/CREB/BDNF pathway in rats. Nutrients. 2019;**11**(4):888

[30] Camuso S, La Rosa P, Fiorenza MT, Canterini S. Pleiotropic effects of BDNF on the cerebellum and hippocampus: Implications for neurodevelopmental disorders. Neurobiology of Disease. 2022;**163**:105606

[31] Singh KK, Park KJ, Hong EJ, Kramer BM, Greenberg ME, Kaplan DR, et al. Developmental axon pruning mediated by BDNF-p75NTR–dependent axon degeneration. Nature Neuroscience. 2008;**11**(6):649-658

[32] Ramachandran B, Frey JU. Interfering with the actin network and its effect on long-term potentiation and synaptic tagging in hippocampal CA1 neurons in slices in vitro. Journal of Neuroscience. 2009;**29**(39):12167-12173

[33] Marini AM, Jiang X, Wu X, Tian F, Zhu D, Okagaki P, et al. Role of brain-derived neurotrophic factor and NF-κB in neuronal plasticity and survival: From genes to phenotype. Restorative Neurology and Neuroscience. 2004;**22**(2):121-130

[34] Xiao D, Liu L, Li Y, Ruan J, Wang H. Licorisoflavan A exerts antidepressant-like effect in mice: Involvement of BDNF-TrkB pathway and AMPA receptors. Neurochemical Research. 2019;**44**:2044-2056

[35] Germanà A, Guerrera MC, Laurà R, Levanti M, Aragona M, Mhalhel K, et al. Expression and localization of BDNF/TrkB system in the zebrafish inner ear. International Journal of Molecular Sciences. 2020;**21**(16):5787

[36] Yang J, Siao C-J, Nagappan G, Marinic T, Jing D, McGrath K, et al. Neuronal release of proBDNF. Nature Neuroscience. 2009;**12**(2):113-115

[37] Shokouhi G, Kosari-Nasab M, Salari A-A. Silymarin sex-dependently improves cognitive functions and alters TNF-α, BDNF, and glutamate in the hippocampus of mice with mild traumatic brain injury. Life Sciences. 2020;**257**:118049

[38] Notaras M, van den Buuse M. Brain-derived neurotrophic factor (BDNF): Novel insights into regulation and genetic variation. The Neuroscientist. 2019;**25**(5):434-454

[39] Hartmann D, Drummond J, Handberg E, Ewell S, Pozzo-Miller L. Multiple approaches to investigate the transport and activity-dependent release of BDNF and their application in neurogenetic disorders. Neural Plasticity. 2012;**20**(12):1-11

[40] Himi N, Takahashi H, Okabe N, Nakamura E, Shiromoto T, Narita K, et al. Exercise in the early stage after stroke enhances hippocampal brain-derived neurotrophic factor expression and memory function recovery. Journal of Stroke and Cerebrovascular Diseases. 2016;**25**(12):2987-2994

[41] Cunha C, Brambilla R, Thomas KL. A simple role for BDNF in learning

and memory? Frontiers in Molecular Neuroscience. 2010;**3**(1):1-14

[42] Nieweg K, Schaller H, Pfrieger FW. Marked differences in cholesterol synthesis between neurons and glial cells from postnatal rats. Journal of Neurochemistry. 2009;**109**(1):125-134

[43] Gaitán JM, Moon HY, Stremlau M, Dubal DB, Cook DB, Okonkwo OC, et al. Effects of aerobic exercise training on systemic biomarkers and cognition in late middle-aged adults at risk for Alzheimer's disease. Frontiers in Endocrinology. 2021;**12**:660181

[44] Enzinger C, Dawes H, Johansen-Berg H, Wade D, Bogdanovic M, Collett J, et al. Brain activity changes associated with treadmill training after stroke. Stroke. 2009;**40**(7):2460-2467

[45] Griffin ÉW, Mullally S, Foley C, Warmington SA, O'Mara SM, Kelly ÁM. Aerobic exercise improves hippocampal function and increases BDNF in the serum of young adult males. Physiology & Behavior. 2011;**104**(5):934-941

[46] Jeon YK, Ha CH. The effect of exercise intensity on brain derived neurotrophic factor and memory in adolescents. Environmental Health and Preventive Medicine. 2017;**22**(1):1-6

[47] Mackay CP, Kuys SS, Brauer SG. The effect of aerobic exercise on brain-derived neurotrophic factor in people with neurological disorders: A systematic review and meta-analysis. Neural Plasticity. 2017;**20**(1):1-19

[48] El-Tamawy MS, Abd-Allah F, Ahmed SM, Darwish MH, Khalifa HA. Aerobic exercises enhance cognitive functions and brain derived neurotrophic factor in ischemic stroke patients. NeuroRehabilitation. 2014;**34**(1):209-213

[49] Dai Y, Ding H, Lu X, Wu X, Xu C, Jiang T, et al. CCRT and aerobic exercise: A randomised controlled study of processing speed, cognitive flexibility, and serum BDNF expression in schizophrenia. Schizophrenia. 2022;**8**(1):84

[50] Lippi G, Mattiuzzi C, Sanchis-Gomar F. Updated overview on interplay between physical exercise, neurotrophins, and cognitive function in humans. Journal of Sport and Health Science. 2020;**9**(1):74-81

[51] Cassilhas RC, De Sousa RAL, Caxa L, Viana V, Meeusen R, Goncalves FL, et al. Indoor aerobic exercise reduces exposure to pollution, improves cognitive function, and enhances BDNF levels in the elderly. Air Quality, Atmosphere & Health. 2022;**15**(1):35-45

[52] Górna S, Domaszewska K. The effect of endurance training on serum BDNF levels in the chronic post-stroke phase: Current evidence and qualitative systematic review. Journal of Clinical Medicine. 2022;**11**(12):3556

[53] Wang D-C, Lin H-T, Lee Y-J, Yu H-F, Wu S-R, Qamar MU. Recovery of BDNF and CB1R in the prefrontal cortex underlying improvement of working memory in prenatal DEHP-exposed male rats after aerobic exercise. International Journal of Molecular Sciences. 2020;**21**(11):3867

[54] Babaei P, Azari HB. Exercise training improves memory performance in older adults: A narrative review of evidence and possible mechanisms. Frontiers in Human Neuroscience. 2022;**15**:771553

[55] Cho HC, Kim JK, Lee NJ, Kim SY, Yoon NK. Effects of combined exercise on cardiovascular risk factors and serum bdnf level in mid-aged women. Journal of Exercise Nutrition & Biochemistry. 2014;**18**(1):61

[56] Aktitiz S, Atakan MM, Turnagöl HH, Koşar ŞN. Interleukin-6, undercarboxylated osteocalcin, and brain-derived neurotrophic factor responses to single and repeated sessions of high-intensity interval exercise. Peptides. 2022;**157**:170864

[57] Schuch FB, da Silveira LE, de Zeni TC, da Silva DP, Wollenhaupt-Aguiar B, Ferrari P, et al. Effects of a single bout of maximal aerobic exercise on BDNF in bipolar disorder: A gender-based response. Psychiatry Research. 2015;**229**(1-2):57-62

[58] Luo L, Li C, Du X, Shi Q, Huang Q, Xu X, et al. Effect of aerobic exercise on BDNF/proBDNF expression in the ischemic hippocampus and depression recovery of rats after stroke. Behavioural Brain Research. 2019;**362**:323-331

[59] Alcantara CC, García-Salazar LF, Silva-Couto MA, Santos GL, Reisman DS, Russo TL. Post-stroke BDNF concentration changes following physical exercise: A systematic review. Frontiers in Neurology. 2018;**9**:637

[60] Sleijser-Koehorst ML, Koop MA, Coppieters MW, Lutke Schipholt IJ, Radisic N, Hooijmans CR, et al. The effects of aerobic exercise on neuroimmune responses in animals with traumatic peripheral nerve injury: A systematic review with meta-analyses. Journal of Neuroinflammation. 2023;**20**(1):1-27

[61] Cheng S-M, Lee S-D. Exercise training enhances BDNF/TrkB signaling pathway and inhibits apoptosis in diabetic cerebral cortex. International Journal of Molecular Sciences. 2022;**23**(12):6740

[62] El-Shamarka ME-S, El-Sahar AE, Saad MA, Assaf N, Sayed RH. Inosine attenuates 3-nitropropionic acid-induced Huntington's disease-like symptoms in rats via the activation of the A2AR/BDNF/TrKB/ERK/CREB signaling pathway. Life Sciences. 2022;**300**:120569

[63] Pinho RA, Aguiar AS Jr, Radák Z. Effects of resistance exercise on cerebral redox regulation and cognition: An interplay between muscle and brain. Antioxidants. 2019;**8**(11):529

[64] Lira FS, de Freitas MC, Gerosa-Neto J, Cholewa JM, Rossi FE. Comparison between full-body vs. split-body resistance exercise on the brain-derived neurotrophic factor and immunometabolic response. The Journal of Strength & Conditioning Research. 2020;**34**(11):3094-3102

[65] Lodo L, Moreira A, Bacurau RFP, Capitani CD, Barbosa WP, Massa M, et al. Resistance exercise intensity does not influence neurotrophic factors response in equated volume schemes. Journal of Human Kinetics. 2020;**74**:227

[66] Quiles JM, Klemp A, Dolan C, Maharaj A, Huang C-J, Khamoui AV, et al. Impact of resistance training program configuration on the circulating brain-derived neurotrophic factor response. Applied Physiology, Nutrition, and Metabolism. 2020;**45**(6):667-674

[67] Fujimura H, Altar CA, Chen R, Nakamura T, Nakahashi T, Kambayashi J-I, et al. Brain-derived neurotrophic factor is stored in human platelets and released by agonist stimulation. Thrombosis and Haemostasis. 2002;**87**(04):728-734

[68] Church DD, Hoffman JR, Mangine GT, Jajtner AR, Townsend JR, Beyer KS, et al. Comparison of high-intensity vs. high-volume resistance training on the BDNF response to exercise. Journal of Applied Physiology. 2016;**121**(1):123-128

[69] de Almeida AA, Gomes da Silva S, Lopim GM, Vannucci Campos D, Fernandes J, Cabral FR, et al. Resistance exercise reduces seizure occurrence, attenuates memory deficits and restores BDNF signaling in rats with chronic epilepsy. Neurochemical Research. 2017;**42**:1230-1239

[70] Zhang H, Liang J-L, Wu Q-Y, Li J-X, Liu Y, Wu L-W, et al. Swimming suppresses cognitive decline of HFD-induced obese mice through reversing hippocampal inflammation, insulin resistance, and BDNF level. Nutrients. 2022;**14**(12):2432

[71] Liu Y, Chu JMT, Ran Y, Zhang Y, Chang RCC, Wong GTC. Prehabilitative resistance exercise reduces neuroinflammation and improves mitochondrial health in aged mice with perioperative neurocognitive disorders. Journal of Neuroinflammation. 2022;**19**(1):150

[72] Deus LA, Corrêa HL, RVP N, Reis AL, Honorato FS, Silva VL, et al. Are resistance training-induced BDNF in hemodialysis patients associated with depressive symptoms, quality of life, antioxidant capacity, and muscle strength? An insight for the muscle–brain–renal axis. International Journal of Environmental Research and Public Health. 2021;**18**(21):11299

[73] Arazi H, Dadvand SS, Suzuki K. Effects of exercise training on depression and anxiety with changing neurotransmitters in methamphetamine long term abusers: A narrative review. Biomedical Human Kinetics. 2022;**14**(1):117-126

[74] Chang S-W, Lee H-C. Vitamin D and health—The missing vitamin in humans. Pediatrics & Neonatology. 2019;**60**(3):237-244

[75] Kazemi F, Babri S, Keyhanmehr P, Farid-Habibi M, Rad SN, Farajdokht F. Maternal vitamin D supplementation and treadmill exercise attenuated vitamin D deficiency-induced anxiety- and depressive-like behaviors in adult male offspring rats. Nutritional Neuroscience. 2023;**26**(6):470-482

[76] Alqudah M, Khanfar M, Alfaqih MA, Al-Shboul O, Al-U'datt DAG, Al-Dwairi A, et al. Correlation between vitamin D and serum brain derived neurotropic factor levels in type 2 diabetes mellitus patients. Biomedical Reports. 2022;**16**(6):1-6

[77] Charoenngam N, Shirvani A, Holick MF. Vitamin D for skeletal and non-skeletal health: What we should know. Journal of Clinical Orthopaedics and Trauma. 2019;**10**(6):1082-1093

[78] Panahpour H, Ghaheri S, Saadati H. Calcitriol supplementation attenuates cisplatin-induced behavioral and cognitive impairments through up-regulation of BDNF in male rats. Brain Research Bulletin. 2022;**181**:21-29

[79] Rastegar-Moghaddam SH, Hosseini M, Alipour F, Rajabian A, Ebrahimzadeh Bideskan A. The effects of vitamin D on learning and memory of hypothyroid juvenile rats and brain tissue acetylcholinesterase activity and oxidative stress indicators. Naunyn-Schmiedeberg's Archives of Pharmacology. 2022;**395**(3):337-351

[80] Abiri B, Sarbakhsh P, Vafa M. Randomized study of the effects of vitamin D and/or magnesium supplementation on mood, serum levels of BDNF, inflammation, and SIRT1 in obese women with mild to moderate depressive symptoms. Nutritional Neuroscience. 2022;**25**(10):2123-2135

[81] Chung C, Silwal P, Kim I, Modlin RL, Jo E-K. Vitamin D-cathelicidin axis: at the

crossroads between protective immunity and pathological inflammation during infection. Immune Network. 2020;**20**(2)

[82] Antunes BM, Rossi FE, Teixeira AM, Lira FS. Short-time high-intensity exercise increases peripheral BDNF in a physical fitness-dependent way in healthy men. European Journal of Sport Science. 2020;**20**(1):43-50

[83] Casseb GA, Kaster MP, Rodrigues AL. Potential role of vitamin D for the management of depression and anxiety. CNS Drugs. 1 Jul 2019;**33**(7):619-637

[84] Horn GP, Kesler RM, Motl RW, Hsiao-Wecksler ET, Klaren RE, Ensari I, et al. Physiological responses to simulated firefighter exercise protocols in varying environments. Ergonomics. 2015;**58**(6):1012-1021

[85] Bahmani E, Hoseini R, Amiri E. Home-based aerobic training and vitamin D improve neurotrophins and inflammatory biomarkers in MS patients. Multiple Sclerosis and Related Disorders. 2022;**60**:103693

[86] Babaei P, Shirkouhi SG, Hosseini R, Soltani Tehrani B. Vitamin D is associated with metabotropic but not neurotrophic effects of exercise in ovariectomized rats. Diabetology & Metabolic Syndrome. 2017;**9**:1-9

[87] Lasoń W, Jantas D, Leśkiewicz M, Regulska M, Basta-Kaim A. The vitamin D receptor as a potential target for the treatment of age-related neurodegenerative diseases such as Alzheimer's and Parkinson's diseases: A narrative review. Cell. 2023;**12**(4):660

[88] Farhangi N, Jafari A, Akhtari E, Zarghami Khameneh A. Effects of four-week aerobic training, caloric restriction and vitamin D supplementation on serum BDNF levels and body composition in overweight and obese women. Daneshvar Medicine. 2022;**30**(1):12-26

[89] Muraro EN, Sbardelotto BM, Guareschi ZM, de Almeida W, Dos Santos AS, Grassiolli S, et al. Vitamin D supplementation combined with aerobic physical exercise restores the cell density in hypothalamic nuclei of rats exposed to monosodium glutamate. Clinical Nutrition ESPEN. 2022;**52**:20-27

[90] Kazemi F, Babri S, Keyhanmehr P, Farid-Habibi M, Rad SN, Farajdokht F. Maternal vitamin D supplementation and treadmill exercise attenuated vitamin D deficiency-induced anxiety- and depressive-like behaviors in adult male offspring rats. Nutritional Neuroscience. 2023;**26**(6):470-482

[91] Khairy EY, Attia MM. Protective effects of vitamin D on neurophysiologic alterations in brain aging: Role of brain-derived neurotrophic factor (BDNF). Nutritional Neuroscience. 2021;**24**(8):650-659